Biomechanics II

Medicine and Sport

Vol. 6

Published for and on behalf of the Research Committee, International Council of Sport and Physical Education, UNESCO

University Park Press · Baltimore · London · Tokyo

Biomechanics II

Editors: J. Vredenbregt, Eindhoven, and J. Wartenweiler, Zurich

With 177 figures and 21 tables

University Park Press · Baltimore · London · Tokyo 1971

Originally published by S. Karger AG, Basel, Switzerland
Distributed exclusively in the United States of America and Canada by University Park Press, Baltimore, Maryland (ISBN 0-8391-0530-4)

Library of Congress Catalog Card Number 72-166285

Medicine and Sport

University Park Press distributes all volumes in USA and Canada

Vol. 1 please ask for details
Vol. 2 Biomechanics
Technique of Drawings of Movement and Movement Analysis.
First International Seminar of Biomechanics, Zurich 1967
Edited by J. WARTENWEILER, Zurich; E. JOKL, Lexington, Ky. and M. HEBBELINCK, Brussels
XXIV+352 p., 216 fig., 17 tab., 1968.
Vol. 3 Biochemistry of Exercise
First International Symposium on Exercise Biochemistry, Brussels 1968
Edited by J. R. POORTMANS, Brussels
X+384 p., 127 fig., 83 tab., 1969.
Vol. 4 Physical Activity and Aging
With special reference to the effect of exercise and training on the natural history of arteriosclerotic heart disease
Edited by D. BRUNNER, Tel Aviv-Jaffa and E. JOKL, Lexington, Ky.
XII+315 p., 98 fig., 71 tab., 1970.
Vol. 5 Exercise and Cardiac Death
Edited by E. JOKL and J. T. McCLELLAN, Lexington, Ky.
X+185 p., 37 fig., 8 tab., 1971.

University Park Press, Chamber of Commerce Building,
Baltimore, Maryland 21202

Printed in Switzerland by Buchdruckerei zum Basler Berichthaus AG, Basel
Blocks: Steiner & Co., Basel
Binding by Max Grollimund, Basel

Contents

IV. Kinetics of Human Motion

a) Muscular Properties

b) Principles in Motion

c) Human Motion in Sports

V. Research in Rehabilitation

VI. Techniques and Methods

VII. Miscellaneous

Preface

This volume contains most of the papers presented at the 2nd International Seminar on Biomechanics held at Eindhoven, Netherlands, August 25–29, 1969.

The main theme 'Formation and Organisation of Human Motion' was divided into the following subjects: Integrative Action of the Nervous System, Muscular Coordination, Motor Learning Process, Kinetics of Human Motion, and Rehabilitation. These subjects also appear in the Index and are supplemented by additional sections on Techniques and Methods and Miscellaneous, in order to obtain an adequate classification.

It is gratifying that only a few authors were uncooperative despite our repeated requests for assistance. We hope that the contents of this book may contribute to a better insight into the problems of Human motion both in research and in practical application. In conclusion, we would like to thank Mr. H. J. A. van Beckum for his linguistic help.

J. Vredenbregt
J. Wartenweiler

Introduction

The first Seminar on Biomechanics was held at Zürich, Switzerland, in 1967. The initiative was taken by the International Council of Sport and Physical Education, in particular by Prof. Jokl from the University of Kentucky, Lexington, USA, president of the Research Committee of the ICSPE. Prof. Wartenweiler of the Eidgenössische Technische Hochschule in Zürich, took on the task of organising this first Seminar. It counted 150 participants and was a great success both scientifically and socially.

Mr. Vredenbregt of the Institute for Perception Research at Eindhoven was invited to organise the second Seminar. He acted as chairman of an organising committee in which Prof. Wartenweiler, Zürich, and Prof. Hebbelinck, Brussels, participated.

Biomechanics, the science of the physical aspects of the motion of animal beings, is a subject both baffling by its complexity and intriguing by the interesting problems it raises. It appeals both to scientists and, as Prof. Jokl summed up in the closing session, to those interested in the application of the principle of biomechanics in

1. Exercise and training in sport
2. Traumatology of sport
3. Clinical medicine
4. Exercise in rehabilitation

For this second Symposium the general theme: 'Formation and Organization of Human Motion' was chosen, divided into six special themes, now figuring as chapters in the proceedings.

It was gratifying to welcome 180 participants, the majority of whom came from 23 countries outside the Netherlands.

The secretariat and the organisation was staffed mainly by members of the Institute for Perception Research. Even though this entails a heavy load on the staff of a small laboratory, it is without a doubt that this procedure greatly contributes towards a congenial atmosphere, so vital for the overall success of any symposium. Important additional help was provided by the Technological University, the Travel Office and the Public Relations Department of Philips, as also by Van Lanschot's Bank and the Eindhoven PTT for running a bank counter and a mail counter respectively. Both the Book Exhibition arranged by Messrs. Swets & Zeitlinger, Publishers, and the Scientific Exhibition, towards which many firms contributed, drew considerable interest from the participants.

Regarding the present Proceedings of this Seminar many thanks are due to Mr. VREDENBREGT for the compilation of the manuscripts and to Messrs. Karger, Publishers at Basel, Switzerland, for its publication.

J. F. SCHOUTEN
President of the Seminar

Medicine and Sport, vol. 6: Biomechanics II, pp. 1–27
(Karger, Basel 1971)

Motor Functions of the Human Brain

A Historical Review

E. Jokl, President
Research Committee, ICSPE

The history of our knowledge of motor functions of the human brain encompasses problems of relevance to anatomy, physiology, anthropology, clinical medicine, pathology, and philosophy. Its inclusion in the study of bio-mechanics signifies the latter's distinction from mechanics as it concerns the physicist and the engineer.

Aristotle thought that the heart is 'the seat of the soul'. Two millenia later, Descartes localized the interaction between soul and body at the pineal gland in the brain. Both Aristotle and Descartes had hypothesized their conclusions from observations of human movements. In 1664 Thomas Willis of Oxford pointed out that hysteria is not a uterine affection as its name implies. 'It is not', he wrote. 'It is an affection of the brain.' S. T. Soemmerring, the neurologist, believed that since all cranial nerves terminate in the walls of the cerebral ventricles, the cerebro-spinal fluid must be the medium of motor activity and thus the seat of the soul, an assertion which was rejected by Immanuel Kant, the philosopher.

In the early years of the 19th century, F. J. Gall and J. C. Spurzheim presented an elaborate theory of cerebral localization, referring to it as phrenology, a pseudo-science derived from the assertion that the brain was the organ of the mind made up of multiple sub-units upon which feeling, sentiments, and propensities, as well as intellectual faculties depend. GALL and SPURZHEIM claimed that each of these had a separate localization that could be charted precisely on the skull. Among them, they said, were such features as amativeness, precociousness, sublimity, and ideality.

The lay public was much impressed by the claims of the phrenologists even though most physicians considered them absurd. E.G. FLOURENS of Paris looked upon the brain as a 'sensorium commune', basing his views upon results he had obtained during crude experiments in which after exposure of the cerebrum of dogs and birds, he produced muscular contractions 'by pinching the brain stem'.

The great breakthrough came in 1861 when PIERRE PAUL BROCA demonstrated before the Anthropological Society of Paris the brain of a patient who had suffered loss of the faculty of speech but not a paralysis of the muscles employed in phonetics. At autopsy, a lesion in the left frontal lobe of the brain was found. From BROCA's demonstration originated a new epoch of neurophysiology and clinical neurology.

Another milestone in the history of localization of motor representation in the brain was reached in 1870 when HITZIG and FRITSCH of Berlin discovered that the cerebral cortex of dogs is electrically excitable. They found that the anterior region of the roof brain gave rise to motor responses but that the posterior parts were inexcitable. They also were the first to observe that stimulation of one side of the brain elicited muscular contraction on the opposite side of the body. In 1874 R.R. BARTHOLOW of Cincinnati, Ohio, applying weak faradic current to the cortex of a girl who had cancer of the skull, produced movements in the contralateral side of the body but also a major epileptic seizure. In 1876 HUGHLINGS JACKSON, pupil of BROWN-SEQUART, became interested in the physiology of the brain cortex when his wife developed cerebral thrombo-phlebitis from which she died. During her illness she had a number of epileptic seizures of the type now referred to as 'Jacksonian', a term first suggested by CHARCOT. During the last quarter of the 19th century, knowledge of motor control areas of the brain was advanced by F.L. GOLTZ of Strassburg and D. FERRIER of London, both of whom were teachers of SHERRINGTON. In 1901 the latter, together with A.S.F. GRÜNBAUM, presented before the Royal Society the results of a detailed study of the motor area of the roof brain of the chimpanzee.

Prior to World War II, four different methods had been elaborated for the identification of motor areas of the human brain: analyses of motor accompaniments of epileptic seizures; electrical stimulation of the surface of the human brain during surgical operations; histological studies of normal and abnormal cerebral tissue; and classification of electroencephalographic patterns obtained from different parts of the brain.

One of the clinical centers in which the cortical epilepsies were studied over many years was the Neurological Institute at McGill University in Montreal under the direction of PENFIELD. He and his colleagues contributed also to the advancement of clinical electroencephalography. PENFIELD's interest in electrical stimulation of the brain during surgical operations had been aroused during his stay in the early thirties at the Neuro-Surgical Clinic in Breslau with OTFRID FOERSTER. In 1936 FOERSTER published a monograph entitled 'Motorische Felder und Bahnen', which has remained the standard work on the subject. The first complete histological analysis of the human cerebral cortex was undertaken before 1920 by BRODMAN and the two VOGTS. However, it took another fifty years before evidence was available from which the relationship of morphological and functional characteristics of the roof brain could be assessed. A hypothesis proposed in 1930 by KLEIST that the two 'brain maps' are identical could not be substantiated.

With the death of FOERSTER in 1945 and of SHERRINGTON in 1952 there came to an end a classical period of research in neurology that was characterized by a preponderance of analytical approaches. Though in his final work, 'Man on his nature', completed in the 90th year of his life, SHERRINGTON referred to 'the motor act as a holistic entity', all his experimental contributions as well as those of his pupils, chief among them those of RUDOLF MAGNUS, had been concerned but with 'fragments of motor behavior'.

During the past 25 years, three developments have been initiated which resulted in a categorical reorientation of scientific concepts of the human nervous system as well as of man and his nature. The first of these new developments was the clarification of the role played in the control of human movements by large regions of the cerebral cortex, e.g. the parietal lobes on which MACDONALD CRITCHLEY of London wrote a noteworthy monograph; secondly, the discovery of the relevance of the limbic system, also known as the visceral brain as well as the rhinencephalon whose relationship to the temporal lobe has lately been summarized by DENIS WILLIAMS and by DIETER JANZ; and thirdly, the elaboration of the phenomenological view of human posture and movement, chiefly by ERWIN STRAUS and F. J. J. BUYTENDIJK.

The significance of the latter development lies in the fact that it considers human movements to be occurrences experienced by the individual as total events that take place in the macroscopic world. In the light of such a view, the important researches referred to before, i.e. those conducted prior to the beginning of the second half of this century, must be considered but building stones of 'theaters of the events', to use a phrase coined by the 17th-century French philosopher JEAN FERNEL; while 'the event itself' is man in action, dancing or playing or working, as well as man reflecting upon what he is doing.

As far as the study of motor functions of the human brain is concerned with the design as well as with the experience of kinetic processes that are of relevance to the student of biomechanics, their scientific and philosophical elaboration are bound to result in a deeper understanding of the theme of the Second International Seminar on Biomechanics, held under the auspices of the UNESCO International Council of Sport and Physical Education at Eindhoven.

Biographical Data

The history of our knowledge of the motor functions of the human brain can be divided into three eras. The first started with DESCARTES (1596–1650) and ended with FLOURENS (1794–1867). It was characterized by efforts to formulate philosophical interpretations of the functions of the central nervous system by means of theories which went beyond observable facts. A second era began in 1861 with the demonstration by P.P. BROCA (1824–1880) that the human brain contains localizable areas to which specific functional significance is attached. It ended with the death in 1941 of OTFRID FOERSTER (born 1873). During this period, an enormous array of data were derived from anatomical, physiological, and clinical observations, while philosophical evaluation of the evidence was on the whole considered 'unscientific'. The latter statement, however, represents a generalization whose validity cannot be fully sustained, more especially in view of the fact that prior to the turn of the century two distinguished British neurologists, HUGHLINGS JACKSON (1835–1911) and CHARLES SHERRINGTON (1857–1952), anticipated the leitmotif of the third era in which we now live. The breakthrough of this era was caused by the teachings of HUSSERL, the philosopher. Their relevance to the understanding of man in general and of human movements in particular was elaborated by a group of philosophically schooled physiologists and clinicians led by F. J. J. BUYTENDIJK and ERWIN STRAUS.

JACKSON's teachings – not fully appreciated during his lifetime – that neurophysiology and clinical neurology reveal characteristics of their own which cannot be derived from pure physics and mathematics; and SHERRINGTON's subsequent statement that individual mind can communicate with other minds solely through movements including speech, proved themselves keys also to the understanding of biomechanics. They left no doubt that biomechanics is a branch of physiology and not of engineering *sensu strictiori*. 'The muscle', SHERRINGTON wrote, 'is the cradle of recognizable mind'. Engineering, to which bio-mechanics is so much indebted, has decisively contributed to the understanding of structure and function of muscle; but it is inherently incapable of doing likewise in respect of mind.

Portraits

1. RENÉ DESCARTES (1596–1650) developed a dualistic philosophy based on the assumption that the mind is separated from the body. The body, he assumed, was constructed like a machine and directed by the rational soul which, according to him, communicated with the mind — 'res cogitans' — in the pineal body in the brain. Messages between body and mind, he wrote, traveled by way of the nerves which carried animal spirits into the muscles, causing them to move as 'flowing water moves the puppets and figures in the grottos and fountain ornaments in our king's gardens'.

2. In his book *Cerebri anatome* THOMAS WILLIS (1621–1675) asserted that voluntary movements are directed from the cerebral hemispheres, involuntary movements from the cerebellum, alluding in this context for the first time to sensory and motor components, of reflex action. He thus established a concept whose validity was subsequently elaborated through experimental studies and clinical observations. WILLIS believed that thought takes place in the corpus callosum, that memories are stored in the cerebral cortex, and that the corpus striatum receives all sensations and is therefore the seat of the 'sensus communis'. His ideas were subsequently elaborated by FLOURENS.

3. FRANZ JOSEF GALL (1758–1828) was born in Germany where he acquired early in his career solid knowledge of neuroanatomy. Later on, he put forward the idea that the brain was the organ of mind and made up of multiple entities representing such characteristics as 'amativeness', 'philoprogenitiveness', and 'inhabitiveness'. Together with his friend, J. C. SPURZHEIM, he founded the pseudoscience of 'phrenology', asserting that each of the cerebral components under review represented a center of mental or moral activity, and that its hypertrophy or atrophy could be detected by palpation of the skull. In retrospect the significance of Gall's fantastic theories appears to have been that he insisted upon functional entities to be localized in the human brain, among them one which was subsequently proved correct by Broca, namely that a center for the motor control of speech is situated in the left frontal lobe.

4. During his lifetime, Pierre Flourens (1794–1867) was considered the chief opponent of Gall and Spurzheim whose 'phrenology' he discredited. Elaborating Willis' teaching that the brain represents a sensorium commune, Flourens held that the cerebrum functions as an indivisible entity subserving the special senses, the intellectual faculties as well as bodily movements.

5. On April 18, 1861, Pierre Paul Broca (1824–1880) presented before the 'Société d'Anthropologie de Paris' a patient who had been unable to speak. He suffered, Broca said, from 'aphémie'. At autopsy this patient's brain showed a lesion of the third frontal cerebral convolution on the left side. Broca's demonstration established the basis from which the theory of brain localization originated.

6. In 1874, at the age of 26, Carl Wernicke (1848–1904) of Breslau published a paper on 'sensory aphasia'. Under certain conditions, he reported, patients were unable to understand spoken words although they heard well and could discriminate articulated sound. Also, they could vocalize words. To explain these observations Wernicke postulated the existence of separate sensory and motor centers of speech in the brain. Autopsy results suggested that a sensory speech center was situated in the posterior part of the first convolution of the left temporal lobe of the brain. If a cerebral lesion happens to be extensive, it may damage both the motor and sensory centers of speech. When in 1905 Pierre Marie (1853–1940) re-examined the brain Broca had presented in 1861 to the 'Société d'Anthropologie', he found that the lesion which affected the motor center of speech included the first convolution of the left temporal lobe as well.

Wernicke's paper gave rise to many diverse though fruitful interpretations. It led to a number of studies which resulted in the identification of focal regions of the brain from which content and form not only of speech but also of other kinds of human movements may be assumed to be programmed and executed.

7. John Hughlings Jackson's (1835–1911) early neurological thinking was influenced by the writings of Herbert Spencer (1820–1903), the philosopher. From 1862–1911 Jackson worked at the National Hospital for Paralysis and Epilepsy in London whose most brilliant staff member he was destined to become. Of special relevance to the issue under review is Jackson's description of a kind of epileptic seizure beginning with unilateral muscular convulsions whose distribution, he pointed out, revealed their origin in the motor cortex. Charcot of Paris applied to the syndrome the term, 'Jacksonian epilepsy'. Although Jackson postulated cortical representation of function, he did not suggest a fixed mosaic pattern as Broca had done. However, Jackson took Broca's idea a step forward by asking, 'If *one* function, speech, can be localized in the brain, why not more than one?'.

8. EDUARD HITZIG (1838–1907) of Berlin corroborated the validity of HUGHLINGS JACKSON's theory derived from clinical observations that the structural design of the motor cortex of man reflects distinct functional patterns. Together with G. T. FRITSCH (1838–1927), HITZIG showed that electrical stimulation of the cerebral cortex of dogs induces motor responses. The experiments, carried out in HITZIG's bedroom, consisted of stimulation of the exposed cerebral cortex with galvanic current whose strength he judged appropriate when on application to his tongue the electrodes caused a sensation.

9. In the early nineties CHARLES SHERRINGTON (1857–1952) and ALBERT SIDNEY FRANKAU GRÜNBAUM (1869–1921) extended the scope of the experiments of FRITSCH and HITZIG, whose findings had been corroborated by DAVID FERRIER (1843–1928). Applying unipolar faradization to the cerebral cortex of lightly anesthetized orangs, gorillas, and chimpanzees, SHERRINGTON and GRÜNBAUM demonstrated that the motor fields of the frontal lobes of anthropoid apes are functionally structured in a manner depicted in figure 3 which they showed at a meeting of the Royal Society in London on November 21, 1901.

10. RUDOLPH MAGNUS (1873–1927), distinguished pupil of SHERRINGTON, undertook a systematic study of 'Haltungs-' and 'Stellreflexe', showing how they are mediated through the central nervous system.

With remarkable experimental skill, sustained industry, and penetrating intelligence, MAGNUS created a new insight into a fundamental aspect of motor control by the brain. Shortly before his death, he completed the manuscript for his classical book, *Körperstellung*.

11. K. BRODMANN (1868–1918), pupil of C. (1875–1962) and O. VOGT (1870–1959), is remembered as the originator of the 'brain map' (fig. 4) published in 1907. In it 52 cerebral cortical areas are shown, each of them identified histologically, BRODMANN's expectation that his cyto-architectonic system of organization of the cerebrum is paralleled by functional equivalents has not been fulfilled. BRODMANN's work, nevertheless, ranks among the great achievements of neurology. Among his findings that have retained special relevance to the issue under study is that of the histological specificity of the structure of the motor cortex of man.

12. HANS BERGER (1873–1941) who in 1919 succeeded OTTO BINSWANGER as Professor of Neuropsychiatry in Jena is best known for his demonstration that action potentials of the human brain can be recorded through the intact cranium. BERGER recognized the epoch-making significance of his observations from which one of the most important methods of investigation of the brain of living man were derived.

'Ich glaube', he wrote in his original paper in 1929, 'das Elektro-Encephalogramm des Menschen gefunden und zum ersten Mal veröffentlicht zu haben'.

13–14. HARVEY CUSHING (1864–1939) and OTFRID FOERSTER (1873–1944) are considered the founders of modern neurosurgery. While working as a young doctor in England and on the European continent, CUSHING came under the influence of

SHERRINGTON in Liverpool and of KOCHER in Berne. After his return to the United States he concentrated on neurological surgery, first at Johns Hopkins and subsequently at the Peter Bent Brigham Hospital in Boston. The last years of his life he spent at Yale University.

OTFRID FOERSTER was born in Breslau. After spending several years in France and Switzerland he returned in 1899 to his native city where he remained the rest of his life. FOERSTER was the great encyclopedist of neurology. Together with O. BUMKE, he edited the *Handbuch der Neurologie*. While CUSHING confined himself to operative neurosurgery, FOERSTER was an outstanding clinical neurologist as well as a self-taught surgeon. In addition, he occupies a unique place in neuro-physiology. He introduced experimental methods to clarify the functional organization of cerebral cortex, basal ganglia, and mid-brain of man.

15. ERWIN STRAUS, born in Frankfurt in 1891, is one of the leading exponents of phenomenology. Starting out with a comprehensive knowledge of neurophysiology, clinical neurology, and psychiatry, he realized early in his career that current approaches to the study of the nervous system relying predominantly on biophysical and biochemical methods are inherently unable to yield a meaningful image of man. Influenced by the writings of E. HUSSERL, the philosopher, STRAUS pointed out that a true anthropology presupposes recognition of human action and experience as unfragmented, holistic phenomena. To the latter, meanings and values are attached of which the exact natural sciences do not even pretend to take cognizance. Thus, he writes, results obtained with quantitatively oriented analytical, microscopic, biochemical, and electrophysical procedures of investigation actually divide 'biomechanical' from 'existential' characteristics of man. STRAUS' monograph, 'Philosophy and Psychiatry' (1969), summarizes much of his concept which, like those of his famous though antithetic forerunner RENÉ DESCARTES, was derived from a consideration of philosophy as 'a tree whose roots are metaphysics, whose trunks are physics, and whose branches are the other sciences'. STRAUS' merit is to have finally shown DESCARTES' classical distinction between a res cogitans and a res extensa to be untenable.

16. WALTER HESS (1881) has been associated with the Zürich Institute of Physiology since 1913. He is the originator of a field of research which has resulted in the clarification of the central control of autonomic accompaniments of all motor acts as well as of a great variety of vegetative phenomena such as sleep, sexual behavior, thermoregulation and sensory integration. Within the context of the survey under study special importance attaches to HESS' investigation of circulatory and respiratory centers in brain stem, diencephalon, thalamus and hypothalamus for which he was awarded a Nobel Prize in 1949. To speak figuratively, every motor act represents a theme which is orchestrated by instruments which the autonomic apparatus controls. The resting pattern of the different components of the autonomic system was described more than a century ago by CLAUDE BERNARD as the 'milieu interne', later by WALTER CANNON as 'homeostasis'. In 1930, JOKL drew attention to the fact that during exercise this homeostatic pattern is deployed according to a different design which he called 'heterostasis'. HESS' most important publications are his monographs 'Die Organisation des vegetativen Nervensystems' (1941); 'Das Zwischenhirn' (1949);

'Diencephalon' (1954); 'Thalamus und Hypothalamus' (1956); and 'Psychologie in biologischer Sicht' (1962).

17. The work of IVAN PETROVITCH PAVLOV (1849–1936) continued the classical tradition of Russian physiology established by I.H. SECHENOV and V.M. BECHTEREV. PAVLOV's thinking was decisively influenced by RUDOLF HEIDENHAIN of Breslau in whose laboratories he conducted important studies in 1890 and 1891. From HEIDENHAIN PAVLOV learned the technique of isolating the pyloric part of the stomach in dogs and the operation resulting in the 'Heidenhain-Pavlov Pouch' which he used in experiments throughout the ensuing four decades. In 1904 PAVLOV received a Nobel Prize 'in recognition of his work on the physiology of digestion'. It was only after 1904 that PAVLOV discovered the 'conditioned reflex' for which he is now best known, first by demonstrating the role played by the cerebral cortex in the control of gastrointestinal function. During the early twenties, PAVLOV extended the scope of his theories by asserting that all motor acts are reflex, either innate or conditioned. The entire cortex, he wrote, is involved in the control of voluntary movements, a view to which he adhered until his death in 1936.

18. SIGMUND FREUD (1856–1939) occupies a unique position among the select group of men who constructed the edifice of our knowledge of the manner in which the human brain controls movements. During the early part of his career, FREUD made noteworthy contributions to physiology, pathology, and clinical neurology. From 1897 onward, however, he confined his efforts to the development of a system of thought which he named 'psychoanalysis'. At the time he was greatly impressed with what he had seen during a visit with J.M. CHARCOT in Paris. In cooperation with his erstwhile friend, JOSEF BREUER, FREUD wrote on borderline aspects of psychiatry, literature, and philosophy: dynamic unconsciousness, transference, resistance, dream analysis, hypnosis, and hysteria, all of them relevant – at times but eccentrically so – to the study of certain facets of form and content of motor behavior. FREUDian concepts proved to be attractive to some – never to all – psychiatrists. Like CHARCOT, FREUD held that 'hysteria was an affection of the mind and not a disease of the nerves'. He taught that both 'nerves and mind were distinctly structured': mind being an 'apparatus', of which the id, the ego, and the super-ego were, he said, constituents. The idea of presenting proof for his assertions never occurred to him.

If one reflects upon FREUD's particular kind of intellectual brilliance as well as upon his lifelong dogmatism, one has to allocate both to psychoanalysis and the treatment derived from it a place outside the realm of clinical medicine. The combination of ritual and arbitrariness which attaches to FREUD's system appears to be a key to the understanding of the fascination which his teachings exert upon disciples, particularly in the United States.

A penetrating comment on FREUD's work was made by SHERRINGTON in a passage which begins with a reference to the fact that more than two millennia ago ARISTOTLE rejected the view that the human brain is 'the seat of the mind'. ARISTOTLE believed that the mind is attached to the heart. Writing in the 1940 edition of his book, 'Man on his Nature', SHERRINGTON expressed himself as follows:

'It might have been expected that ARISTOTLE's attaching mind to what has proved to be a muscular pump instead of to its proper organ would have ruined his description of mental structure and ways. That it does not notably do so is a gloss on macro-anatomy as any clue to mind. A like reflection arises when we read another original authority on mind, one recently taken from us. It is far from ARISTOTLE to FREUD. They are poles asunder. Yet the work of the two has resemblance in one respect. With both of them their study of the mind is wholly divorced from anatomical features of the brain. Even as a background to the metaphors and parables, and classical myths, by means of which psycho-analysis proceeds to tell its story, the anatomy of the brain has no more part there than in the narration of, say, BUNYAN's Pilgrim's Progress.'

FREUD belongs to a small and exceptionally gifted group of physicians who, after completing their medical studies, let their imaginative minds dwell upon literature and philosophy, among them JOHN KEATS, FRIEDRICH SCHILLER, and ANTON CHEKOV.

19. F.J.J. BUYTENDIJK (born 1887), physiologist, anthropologist, and philosopher, ranks as one of the most original and encyclopedic thinkers on human biology of the century. Prior to 1930, BUYTENDIJK had become known as author of numerous studies in experimental physiology, some of them undertaken in collaboration with E.D. ADRIAN in Cambridge. Others were published from his own laboratory in Utrecht, Holland, as well as from physiological institutes in Germany and France. During the 1928 Olympic Games in Amsterdam, BUYTENDIJK organized medical studies whose results he reported in a volume published with Springer, Berlin, in 1929. Early in his career, BUYTENDIJK developed an interest in philosophical interpretation of human motor behavior, first as a result of personal contacts with VIKTOR VON WEIZSÄCKER, E. VON GEBSATTEL, MAX SCHELER, ROMANO GUARDINI, HELMUT PLESSNER, and ERWIN STRAUS. More and more he became convinced that the problem of 'values' – invariably attached to human action – cannot be understood on the basis of biophysical and biochemical analyses. Altogether, he believes, it is the task of phenomenological anthropology to build bridges between scientific medicine and morality.

20. EDGAR DOUGLAS ADRIAN (born 1889), who has been associated with Trinity College, Cambridge, England, since 1908, elucidated the manner in which proprioceptive and other sensory impulses are engendered and transmitted to the central nervous system. In a series of brilliant experiments ADRIAN showed that when muscle is subjected to increased tension, the electric potentials in individual fibers remain unchanged in magnitude but increase in frequency. He subsequently proved that like electric patterns of activation and transfer of stimuli apply to all senses. In a classical monograph 'The Physical Background of Perception' (Oxford, 1947) he demonstrated how 'information' of various kinds reaching the central nervous system is an indispensable prerequisite for the effective execution of voluntary movements, thus establishing the basis for a new understanding of the integrative action of the nervous system, the concept first postulated by SHERRINGTON with whom he shared the Nobel Prize for Medicine in 1932.

1 René Descartes (1596–1650)

2 Thomas Willis (1621–1675)

3 Franz Josef Gall (1758–1828)

4 Pierre Flourens (1794–1867)

5 PIERRE PAUL BROCA (1824–1880)

6 CARL WERNICKE (1848–1904)

7 JOHN H. JACKSON (1835–1911)

8 EDUARD HITZIG (1838–1907)

9 CHARLES SHERRINGTON (1857–1952)

10 RUDOLF MAGNUS (1873–1927)

11 K. BRODMANN (1868–1918)

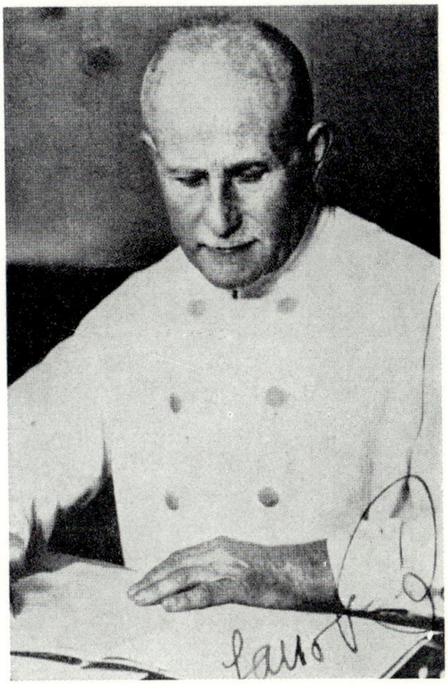

12 HANS BERGER (1873–1941)

13 HARVEY CUSHING (1864–1939)

14 OTFRID FOERSTER (1873–1944)

15 ERWIN STRAUS (born 1891)

16 WALTER HESS (born 1881)

17 Petrovitch Pavlov (1849–1936)

18 Sigmund Freud (1856–1939)

19 F. J. J. Buytendijk (born 1887)

20 E. Douglas Adrian (born 1889)

Historical Illustrations

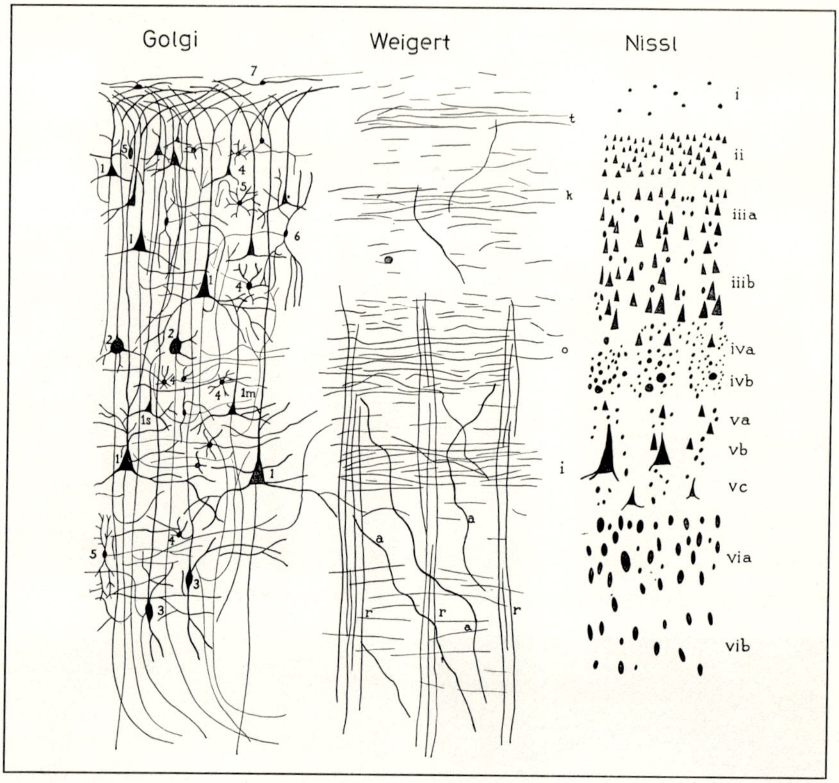

a *b* *c*

Fig. 1. The cerebral cortex exhibits local differences in structure which are generally referred to as 'areas'. These areas and the maps which were drawn up to show their location led many workers to consider the cortex as an aggregate of different organs functioning independently of each other. Areal differences can be ascertained by selective staining methods, chief among them one introduced by NISSL which causes *neurons* to stand out, one by WEIGERT which shows *myelinated fibers,* one by GOLGI which reveals cell bodies, and one by CAJAL who found that pericellular tissue nets, after treatment with a gold chloride-mercuric chloride compound, *d* 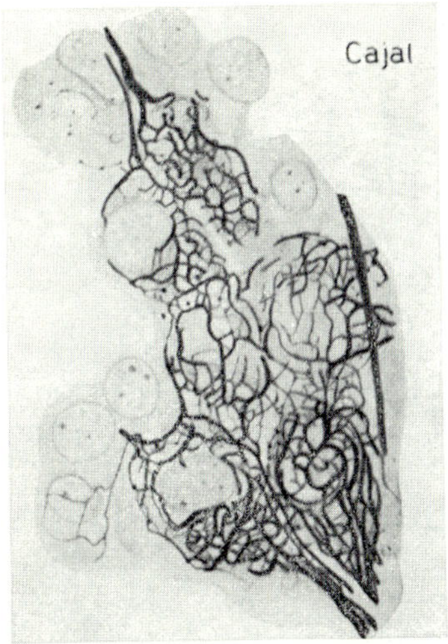

become visible. The *motor cortex* is characterized histologically by exceptionally large cells first described by BETZ. Part of it is called the 'area gigantopyramidalis'.

Scheme of the structure of the cerebral cortex as seen in preparations after GOLGI (outlines of neurons), WEIGERT (myelinated fibers) and NISSL (cell bodies): *1* pyramidal cell, *lm* 'medium pyramid', *ls* 'short pyramid', *2* star pyramid, *3* fusiform cell, *4* star cell, *5* spider cell, *6* double bush cell, *7* horizontal cell of Cajal, *a* oblique fibers, *r* radii, *t* tangential fibers, *k* stripe of Bechterew, *o* outer, *i* inner stripe of Baillarger. The small Roman numerals on the right margin indicate layers (see text).

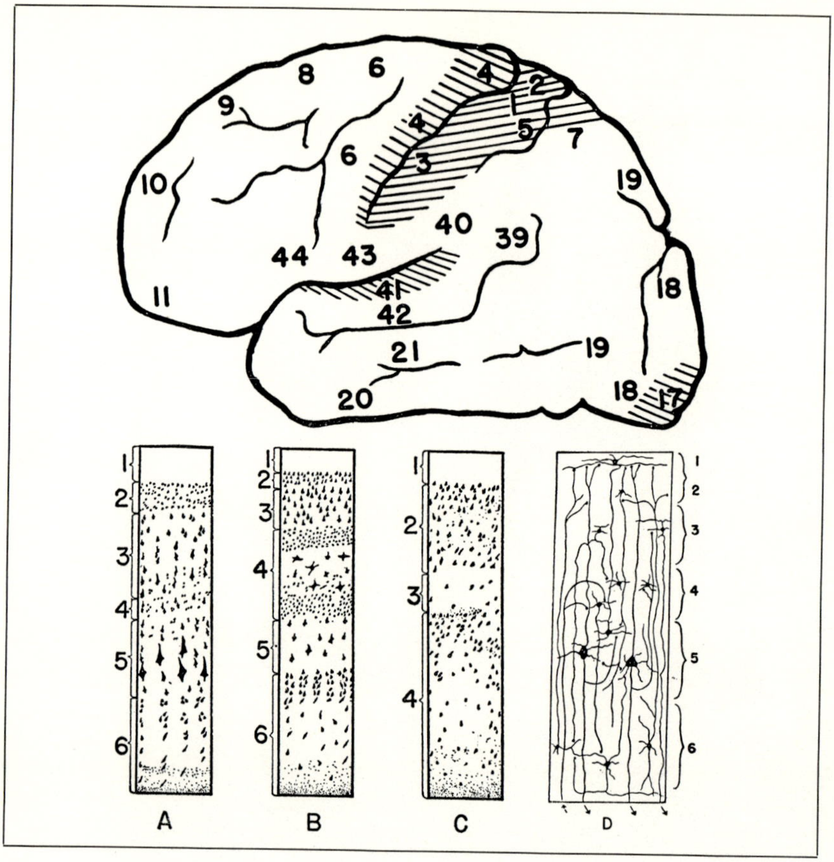

Fig. 2. Lateral aspect of the cerebral cortex showing the principal convolutions and areas. Projection areas are shaded. Below are sketched, in a much simplified manner, four microscopical pictures of sections of gray matter from different areas. 'A' is from area 4 showing the large motor cells in layer 5. Layer 4 is seen to be very narrow in proportion (Nissl stain). 'B' is from area 17 showing the cellular architecture of the visual sensory receiving station. Layer 4 is seen to be much wider, whereas layer 5 is poorly developed; layer 3 is also narrower than in 'A' (Nissl stain). 'C' is from the temporal lobe on its mesial aspect in the olfactory (entorhinal) cortex. This is the old primitive part of the cortex and therefore has only 4 layers (Nissl stain). 'D' picture of area 4 as seen in silver stain (e.g. CAJAL) showing that the Nissl pictures are only a first step in orientation and give a false idea of simplicity, because the axonal connections are complex and mostly radial, whereas the Nissl picture gives the impression of horizontal lamination (after STANLEY COBB).

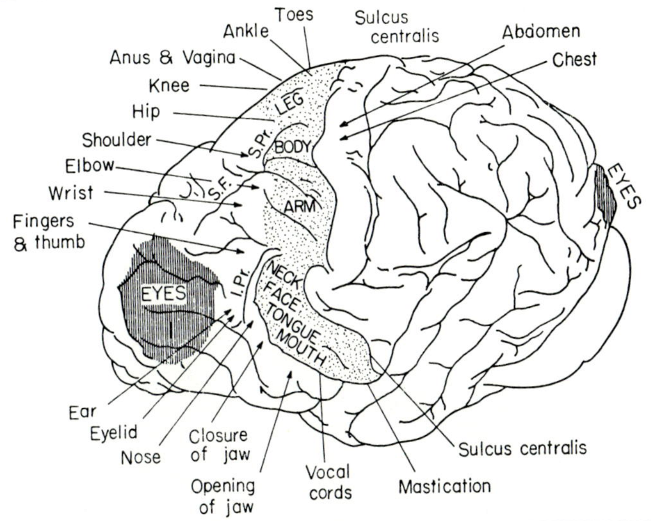

Fig. 3. Cortical motor points (from GRÜNBAUM and SHERRINGTON). Brain of a chimpanzee *(Troglodytes niger)*. Left hemisphere viewed from side and above so as to obtain as far as possible the configuration of the sulcus centralis area. The figure involves, nevertheless, considerable foreshortening about the top and bottom of *sulcus centralis*. The extent of the 'motor' area is hidden in sulci; for instance, the area extends into the *sulc. centralis* and *sulc. precentralis,* also into occasional sulci which cross the precentral gyrus. The names printed large on the stippled area indicate the main regions of the 'motor' area; the names printed small outside the brain, indicate broadly by their pointing lines the relative topography of some of the chief subdivisions of the main regions of the 'motor' cortex. But there exists much overlapping of the areas and of their subdivisions which the diagram does not attempt to show.

The shaded regions, marked EYES, indicate in the frontal and occipital regions respectively the portions of the cortex which, under faradization, yield conjugate movements of the eyeballs. But it is questionable whether these reactions sufficiently resemble those of the 'motor' area to be included with them. They are therefore, marked in vertical shading instead of stippling, used to delineate the 'motor' area.

S.F. Superior frontal sulcus, *S.Pr.* superior precentral sulcus, *I.Pr.* inferior precentral sulcus.

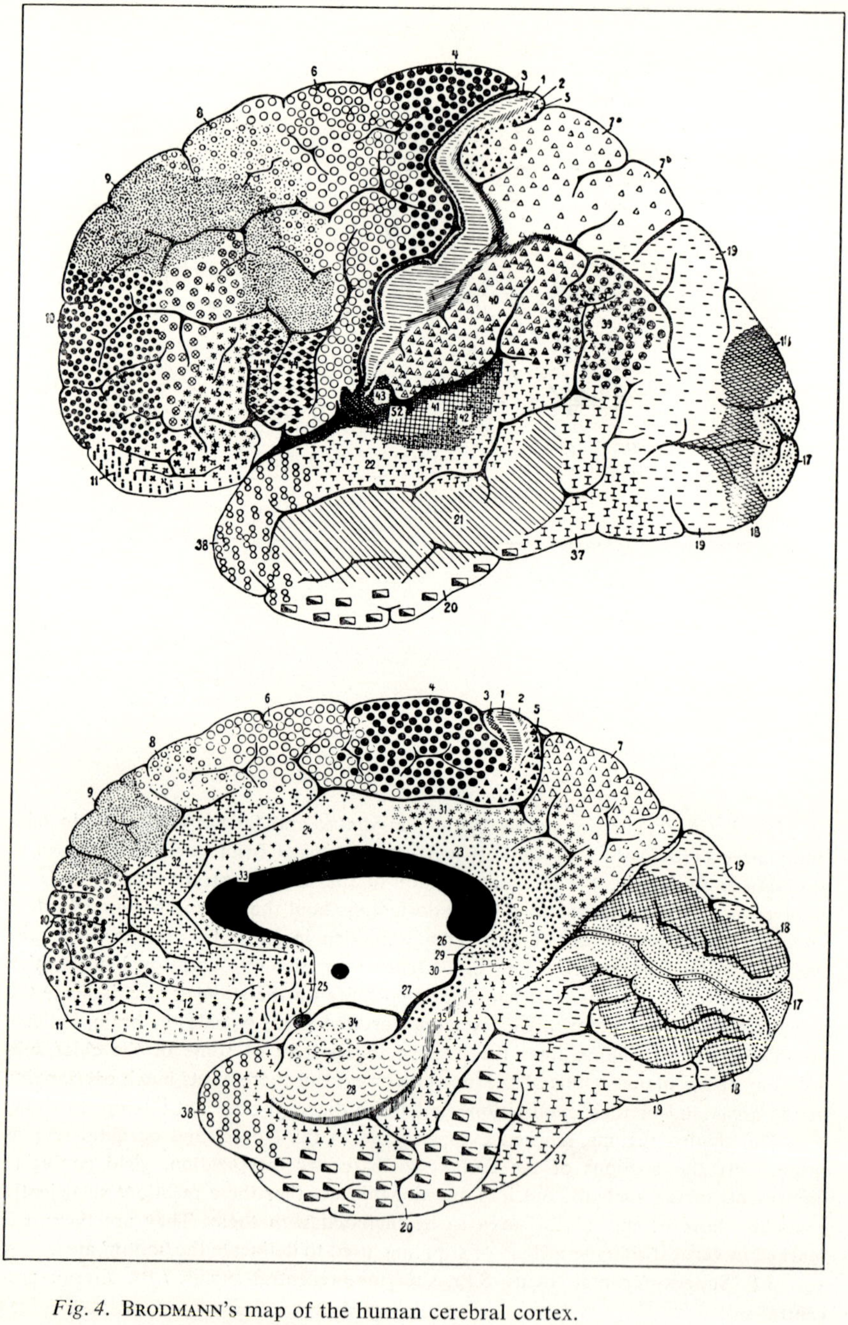

Fig. 4. BRODMANN's map of the human cerebral cortex.

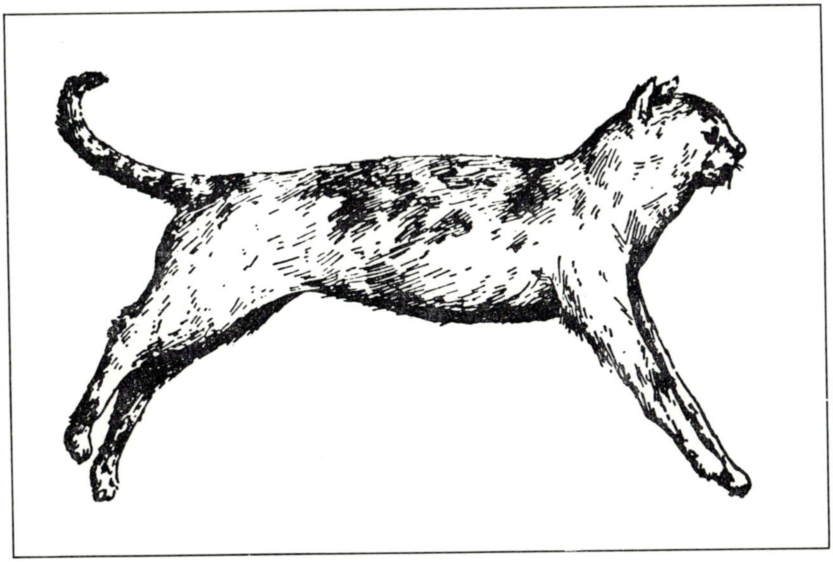

Fig. 5. Enthirnungsstarre ('decerebrate rigidity') (SHERRINGTON).

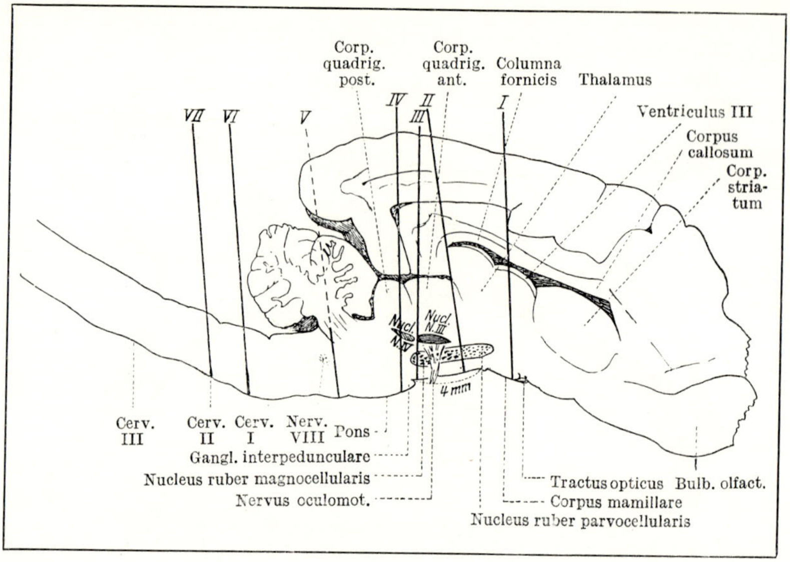

Fig. 6. In 1908 RUDOLF MAGNUS went to Liverpool to work with SHERRINGTON. At that time knowledge of motor control through the brain was restricted to information on the organization of the posterior portion of the frontal cortex as well as to the seemingly inexplicable demonstration in the eighties by SHERRINGTON's teacher, GOLTZ of Strassburg, that dogs after ablation of the 'motor cortex' retain the ability to stand and walk. Furthermore, SHERRINGTON had just rediscovered the fact known before him already to MAGENDIE, FLOURENS, and CLAUDE BERNARD that removal of cerebral hemispheres and brain stem in front of the line marked III in figure 6 causes 'decerebrate rigidity' in cats (fig. 5). Under chloroform narcosis the carotid arteries of cats were tied, the vertebral circulation manually compressed and the brain sectioned between the corpora quadrigemina. A state of extensor rigidity of the profoundly unconscious animal could thus be induced, 'a condition', SHERRINGTON wrote in 1898, 'possessing considerable physiological interest'. In a sustained sequence of brilliant experiments, MAGNUS demonstrated a large variety of postural and attitudinal reflex mechanisms in their dependence upon cerebral structures. In 1924 he described his results in a 740-page book, *Körperstellung,* an opus equivalent in importance to SHERRINGTON's *Integrative Action of the Nervous System.* While working with SHERRINGTON in Liverpool, MAGNUS received the appointment to the Chair of Pharmacology at the University of Utrecht in Holland.

Katze. Linie I: Sprungbereitschaft, Körperstellreflexe auf den Körper.

Linie II: Normale Tonusverteilung, Labyrinthstellreflexe, Liftreaktion.

Linie III: Enthirnungsstarre (Aufhebung der normalen Tonusverteilung und der Labyrinthstellreflexe).

Linie IV: Halsstellreflexe.

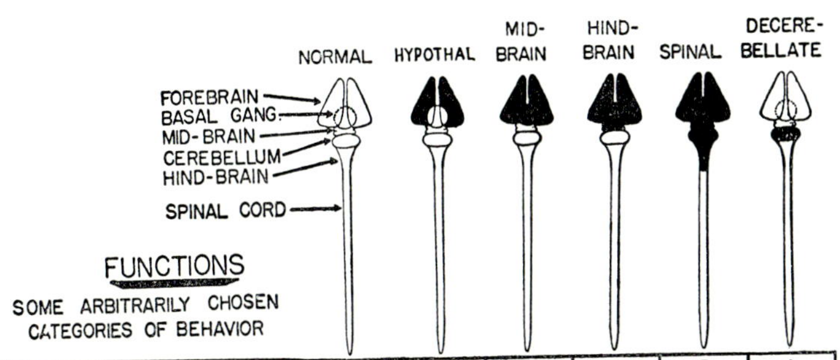

	NORMAL	HYPOTHAL	MID-BRAIN	HIND-BRAIN	SPINAL	DECEREBELLATE
INITIATIVE, ETC.	+	O	O	O	O	+
CONDITIONED REFLEXES	+	O	O	O	O	+
EMOTIONAL "	+	+++	O	O	O	+
LOCOMOTOR "	+	++	+	O	O	IN-COORDINATE
RIGHTING "	+	+	++	O	O	
ANTI-GRAVITY "	+	++	++	+++	O	
RESPIRATIONS	+	+	+	+	O	+
NECK REFLEXES	+	+	+	++	O	+
SPINAL "	+	++	++	++	++	+

Fig. 7. Diagrammatic summary of investigations by many investigators of motor integration in mammals. It is by systematic experimental *disintegration,* through ablation of different parts of the brain, that the data have been collected.

Largely as a result of the work of SHERRINGTON and MAGNUS, the main components involved in the cerebral control of movements can be diagrammatically identified as shown in the above picture. The *normal animal* is in possession of all the functions referred to as initiative, conditioned reflexes, emotional reflexes, locomotor reflexes, righting reflexes, anti-gravity reflexes, respiration, neck reflexes, spinal reflexes. The *hypothalamic* preparation is practically decerebrated, with only the hypothalamus and some structures lateral and caudal to it remaining above the mid-brain. The *mid-brain* preparation has had the cerebral hemispheres and upper basal ganglia removed. The hind-brain preparation is truly decerebrate. The *spinal* preparation has lost all complex behavior, even the power of autonomic integration. The *decerebrate* preparation appears normal at rest (after STANLEY COBB).

Linie V: Tonische Labyrinthreflexe auf die Körpermuskulatur, Kopfdrehreaktion und -nachreaktion, einseitiger Tonusverlust der Glieder nach einseitiger Octavusdurchschneidung.

Linie VI: Tonische Halsreflexe auf die Extremitäten.

Linie VII: (Aufhebung der tonischen Halsreflexe auf die Extremitäten).' (After MAGNUS.)

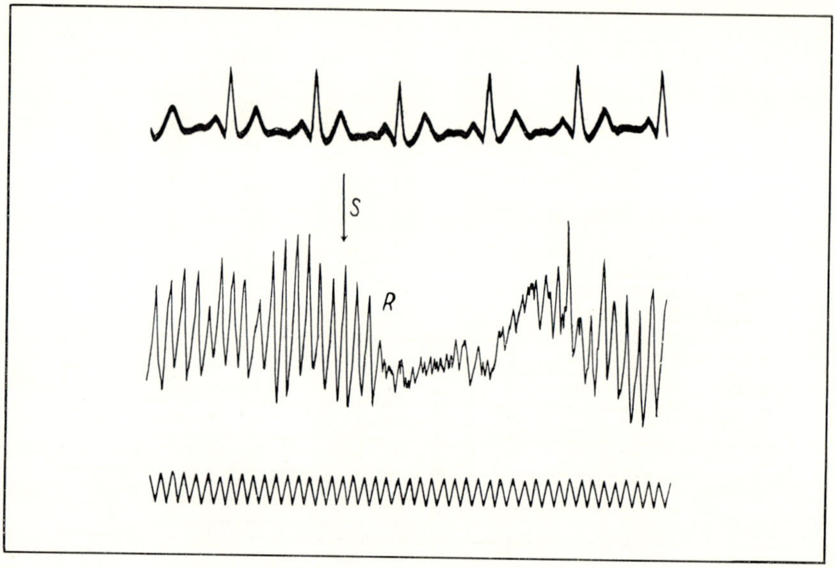

Fig. 8. When in 1929 BERGER presented his epochal findings, he fully realized that electroencephalography would render possible the identification of a variety of accompaniments of human motor acts, of their sensory as well as muscular components. He also anticipated that it would do so under physiological and pathological conditions.

The experimental record published by BERGER in 1935 shows an electroencephalogram (middle), an electrocardiogram (above) and a time marker (below). At point 'S', the subject's left index finger was pricked by a pin; at point 'R', the electroencephalogram responds to the stimulus. Similar studies have since been conducted to determine the electroencephalographic manifestations of *auditory* stimuli such as the sound of the starting pistol at the beginning of a sprint race and to assess the time required for its 'transfer' to the motor cortex of the brain. Electroencephalography as a means of elucidating neurological problems of athletics has received special attention in the Soviet Union. (See article by G. M. KUKOLEWSKIJ, in 'Medizin und Sport' [Berlin, DDR], November 1967, pp. 165–169.)

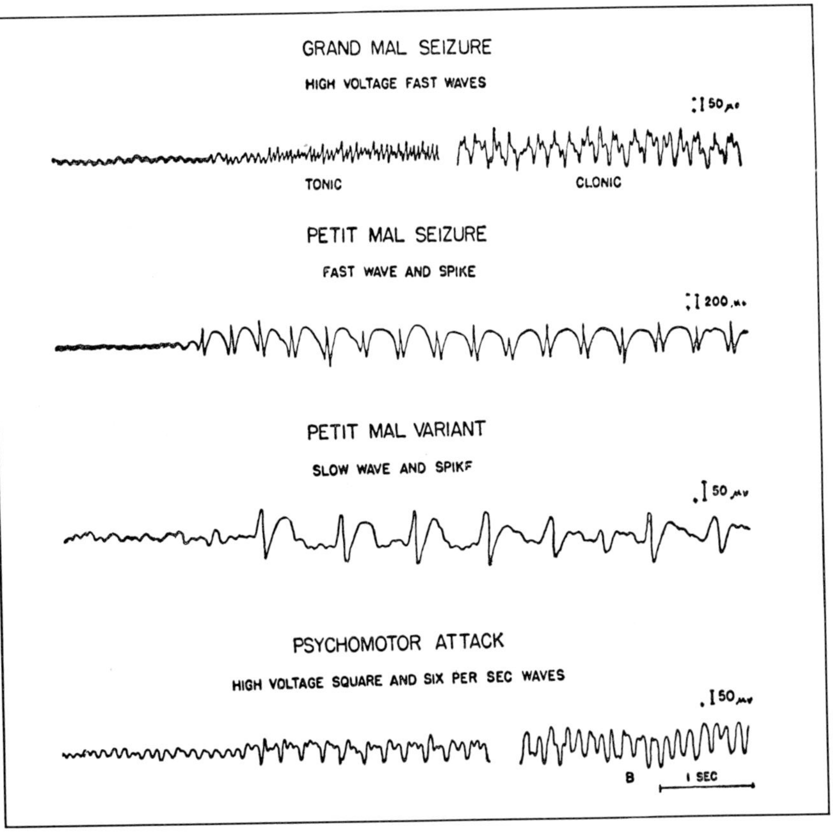

Fig. 9. Representative electroencephalograms of four patients to show records taken before and during four different types of seizures, viz., grand mal, two forms of petit mal, and psychic variant (psychomotor). At the right is the perpendicular deflection made by a 50- or 200-millionth-volt current, and at the bottom the time marked by one second. The left of each tracing is a portion of the person's normal record. The rest of the tracing was made during a seizure. Four different types of seizures are depicted. Uppermost is the tonic and then, after an interval, the clonic phase of a grand mal convulsion. The second tracing is the three-a-second alternate wave-and-spike of a petit mal. The third tracing is of the relatively rare slow (two a second) wave-and-spike called petit mal variant. This is not usually accompanied by symptoms. The fourth tracing was taken during two phases of a psychomotor seizure. The tracings are about one third their natural size. (After W. G. LENNOX.)

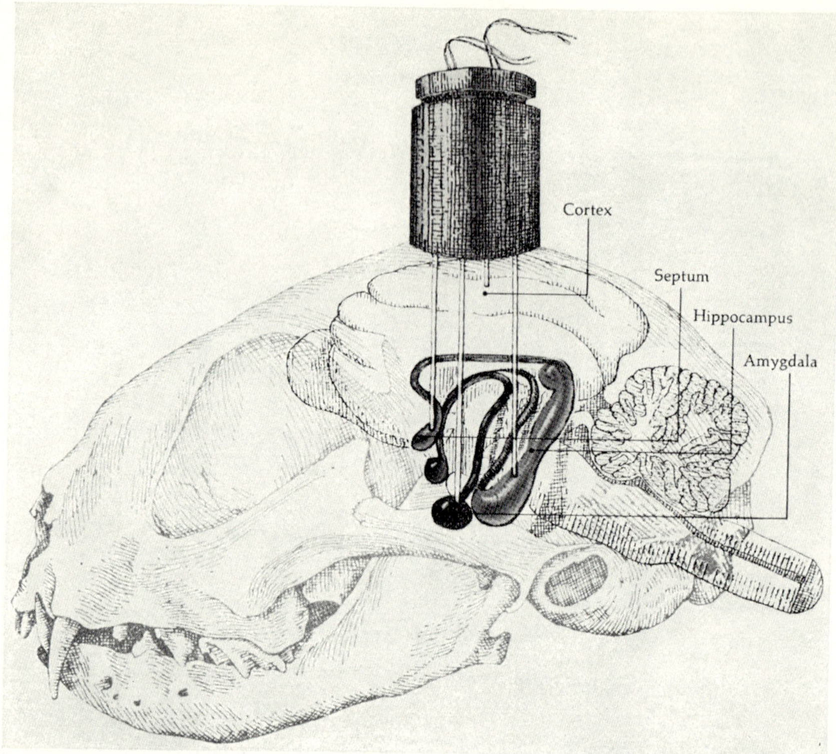

Fig. 10. The limbic system, also known as the visceral brain and the rhinencephalon, plays a fundamental role in *emotional* behavior which includes an important category of *motor acts.* What a person feels is a function of the limbic system; what he thinks, a function of the cortex.

By placing stereo-tactically guided electrodes into thalamus and hypothalamus, W. R. HESS opened up a field of experimental research explored by stimulating or destruction of selected areas. More recently efforts have been made to explore *cortical* and *limbic* components of the *motor control* system of the brain. Motor manifestations and autonomic accompaniments of normal as well as of abnormal emotional states were thus analyzed.

In the above illustration, one electrode is placed into the posterior portion of the frontal cortical area, three others into septum, hippocampus, and amygdala, all of them parts of the limbic system.

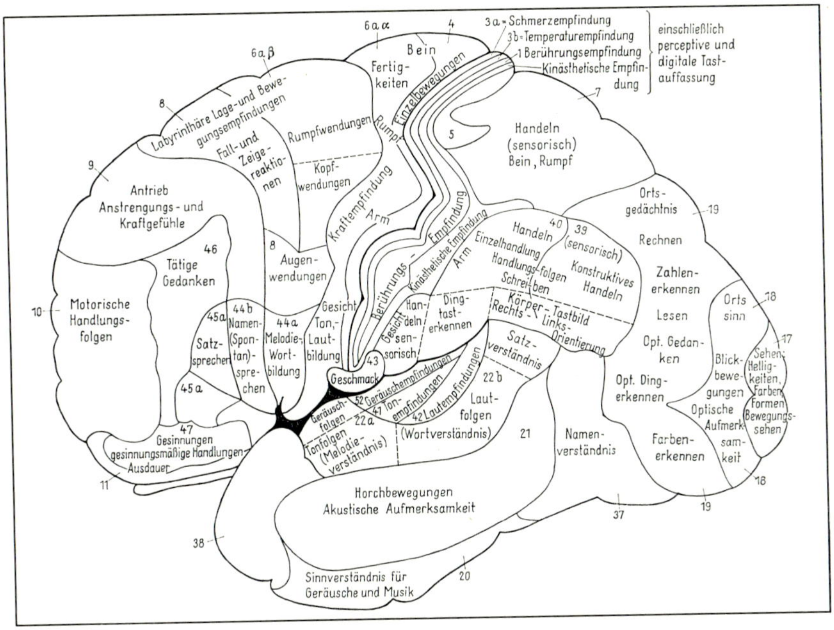

Fig. 11. In 1924 KLEIST of Frankfurt am Main published a 'Hirnkarte' derived from studies of World War I victims with localized brain injuries. KLEIST tried to interpret his clinical and pathological findings in accordance with BRODMANN's cytoarchitectonic field theory. His assumption of a neurophysiological equivalence of the histological differentiation of the cerebral cortex has since been proved untenable.

Author's address: Prof. E. JOKL, University of Kentucky, *Lexington, KY 40506*

I. Integrative Action of the Nervous System

Medicine and Sport, vol. 6: Biomechanics II, pp. 30–33
(Karger, Basel 1971)

Neuromuscular Integration

J. P. SCHADÉ

Central Institute for Brain Research, Amsterdam

The alternation of contraction and relaxation of the muscle is caused by changes in activity of the motor nerves. The motor neurons for the synergic and antagonistic muscles which activate the knee joints are located in the lumbal and sacral part of the spinal cord. Contraction is at the same time accompanied by relaxation. The movement is directed by electrical potentials in the motor nerves which originate from the spinal cord.

Some hundreds of motor neurons innervate the thousands of muscle fibers which form a muscle, and by activation of the motor centers in the brain various phases of contraction and relaxation can be completed. In addition to the motor nerves a large number of sensory nerves transport stimuli from the sensory nerve cells in the periphery of the body towards the spinal cord and brain. Many thousands of sensory nerve cells or receptor cells which are essential to the coordination and regulation of movements are found in the motor system. Without the sensory organs in muscles and tendons, movement would be impossible.

A muscle contains about 50–100 small sensory organs: the muscle spindles. These spindles are of a complex structure and function, and may be considered as extremely small measuring instruments. A muscle spindle is built up from two long striated fibers, somewhat thicker in the middle, and provided with a large number of nuclei. In addition, four short striated fibers are found which have no curvature, and are not contractile in the equatorial part. In the middle the whole is surrounded by a connective tissue membrane, filled with a lymphe-like substance. The muscle spindle contains a double sensory system: around the equatorial parts, the primary system. The secondary system is primarily in contact with the peripheral parts of the spindle. The axons of the primary sensory endings conduct nerve impulses at a velocity of 80 msec, the secondary sensory endings, however, at a velocity of 60 msec. The muscle spindle with its intrafusal fibers and nerves is surrounded by a connective tissue sheath which in turn is surrounded by extrafusal muscle fibers.

How does this regulating mechanism function? The motor neuron with its processes innervates some hundreds of extrafusal muscle fibers, building together an anatomical and functional unit: the motor unit. Together with these motor nerves the sensory nerves of the muscle spindle play an essential role in the regulation of alternating stretch- and contraction movements. The axons of the primary sensory endings of the intrafusal fibers make monosynaptically contact with the motor neurons of the antagonistic muscle – in this case, however, via inhibitory interneurons.

Stretch is an adequate stimulus for the primary sensory endings. Stretch causes a transition of the mechanical stimulus into an electrical signal. The amount of stretch determines the number of electrical potentials. In this way the motor neuron of the muscle is being excited monosynaptically. This monosynaptic excitation causes a depolarization of the neuronal membrane, and electrical signals return to the flexor muscle. Consequently the motor neurons of the extensors are inhibited by the activity of inhibitory interneurons. The number of action potentials of these neurons decreases because the cell is hyperpolarized. At the same time activation of the biceps leads to relaxation of the quadriceps: the quadriceps is now being stretched, and so are the muscle spindles. This causes an increase in the electrical activity of the sensory neurons, and also more efferent potentials, resulting in excitation of the motoneuron of the quadriceps and inhibition of the motoneuron of the biceps; hence, the leg extends.

The alternation of excitation and inhibition, and thus of contraction and relaxation of muscle fibers, occurs with great speed and precision. In addition to muscle spindles which are embedded in the muscle tissue, each muscle contains 50–100 tendon organs. The many fine branchings of these receptors are embedded between the fibers of the tendon. These receptors also respond to stretch. However, the threshold of the sensory nerve fibers of the tendon organs differs markedly from that of the muscle spindles. Only a strong stimulus changes the activity of the tendon organs. Contraction of the muscle causes stretch of the tendons. The tendon organs add an additional dimension to the regulating mechanisms of the motor apparatus.

The sensory fibers of the tendon organ exhibit a different method of action in the spinal cord than those of the muscle spindles. The polysynaptic contact with the motor neurons of the antagonistic and the synergic muscles is seen, but in this circuit an inhibitory interneuron is found. In the neuronal circuit of the antagonistic muscles, on the other hand, only excitatory interneurons are found.

The sensory fibers of the muscle spindles show inhibitory connections with the synergic muscles, and excitatory connections with the antagonistic muscles: contraction of quadriceps – activation of the tendon organs, stretch of the biceps – activation of the muscle spindles; increase of electrical signals in the sensory neurons; inhibition of neurons of the quadriceps muscle. Concommitantly, excitation of the neurons of the biceps is seen via a different neuronal circuit. The quadriceps relaxes, the biceps muscle contracts. The tendon organs system of the quadriceps works jointly with the muscle spindle system in the biceps, and vice versa. Both mechanisms are fully integrated, but the threshold of both systems differs markedly because the muscle spindle mechanism responds to a lower threshold.

One important effect has not been discussed yet: Our attention has been focussed on the moving leg. Seemingly, the left leg does not participate in the movement pattern. However, if this is the case, the rhythmic movement of the right leg would not last very long. The co-ordination of the elements of this complex movement pattern does not occur in the spinal cord, but does occur in the cerebellum. Here functions a fast neuronal computer system which processes all information from the receptor of the motor apparatus. Afferent fibers of the right leg convey electrical impulses to the motor neurons in the spinal cord.

But these signals are also conveyed via ascending tracts to the higher centers of the brain. Information from many parts of the brain (e.g. the vestibular nuclei) converges in the cerebellum, and after processing part of the information is sent back to the spinal cord. These stimuli, however, cause activation of both flexors and extensors of the leg.

Let us now first consider the fiber systems which convey these impulses. The axons of the descending tracts, called the reticulospinal fibers, make primarily synaptic contact with the γ-neurons. These small γ-neurons are located between the α-motor neurons in the spinal cord. The α-neurons with their thick axons innervate the extrafusal fibers. The γ-neurons, having thin axons, innervate the intrafusal fibers of the muscle spindles. In this way the contraction of the intrafusal muscle fibers can be regulated independently. The primary sensory endings of the muscle spindle, however, generally make synaptic contact with the α-cell. When the signals in the descending tracts activate the γ-neurons of both extensors and flexors, the result will be a contraction of the peripheral parts of the intrafusal fibers. The central parts of these muscle fibers are now being stretched, which causes an increase in the electrical signals in the primary sensory endings, which consequently leads to activation of the α-neurons of both flexors and extensors.

Thus the α-neurons of the antagonistic muscles are stimulated in the same way.

An extensive review of this subject appears in 'Progress in Physical Therapy', vol. 1 (1970).

Author's address: Dr. J.P. SCHADÉ, Central Institute for Brain Research. *Amsterdam* (Netherlands)

Medicine and Sport, vol. 6: Biomechanics II, pp. 34–38
(Karger, Basel 1971)

The Cerebral Mechanisms of Automatic Motor Activity and Their Disturbances Induced by Variable Peripheral Troubles

P. Pinelli

Clinica delle Malattie Nervose e Mentali, Università CSC, Roma

Subject Matter

These investigations and studies have been suggested by the clinical observation of cases where complex motor disturbances, commonly considered as psychogenic, are superimposed upon neuromuscular troubles mainly of traumatic origin. This group of disturbances is known as physiopathies of Babinski-Froment or traumatic reflex paralyses or contractures of Purves-Stewart or reflex dystrophy.

The resulting principles to be discussed take on a general value, since the phenomena induced by traumatic lesions here investigated are essentially identical with, among others, the consequences of demyelinating lesions of other origins. Moreover, the repercussions analyzed in motor cortical areas may have the same mechanisms of occurrence as in sensitive areas.

When studying this type of functional consequences in the central nervous system (CNS) it would be erroneous to postulate a principle of proportionality between the magnitude of peripheral functional impairment and the magnitude of the central repercussion. In reality it was observed that the central repercussion does not depend on the intensity of impairment evaluated at clinical level, but on the variability and irregularity of occurrence of the troubles investigated by electrophysiological methods even at subclinical level. On the other hand this absence of proportionality can be partially complicated by the law of non-linear proportionality in the normal functioning of the nervous system (NS) [NAYRAC, 1963].

Terminology

Recent studies on the physical interpretation of mechanisms involved in CNS function have focused attention on some terms introduced by neurologists in the past. However, an arbitrary broadening of meaning has sometimes resulted. The following proposals seem justifiable to me in view of the use of some terms and their neurophysiological implications.

The word *scheme* [RUSSEL BRAIN, 1951] can be used to indicate a functional aggregation of neurons which will assure the material mechanism of a given complex performance rather constant in time.

(1) So the schemes with inherited structure, governing the control of the more constant and unchanging core of the motor act, are localized on basal ganglions connected to the red nucleus and cerebellar system.

(2) Moreover, the existence of acquired schemes of motor acts is to be admitted; they correspond to neuronal assemblies able to assure automatic activities created by preceding facilitations according to the mechanisms discussed by HEBB. The seat of acquired schemes is located in the 6th Broadmann's cortical area and in the supplementary area [PHILLIPS, 1965].

(3) Acquired schemes may be traced back to two sources: (a) the cortical integrating system of MACKAY (also including the gnoso-praxic centres) through a feedforward mechanism; (b) the peripheral afferences occurring as repeated and always identical patterns of impulses ensuring a feedback mechanism at cortical level.

(4) From a comparatively simple motor scheme different *variations in the act* can arise at the moment when the movement is executed. Soon after the impulses for the automatic movement are generated from the scheme, specific actual afferent impulses intervene at various levels [GELLHORN, 1953] to attain the variation most suitable at that very moment. The scheme can accordingly be indicated as functionally *plastic*, since it results from an integrant fringe with multiple choice.

(5) Both voluntary and automatic 'mises en train' have been attributed to a mechanism developed by the centro-encephalic system [PENFIELD *et al.*, 1954] with hypothalamic and amygdalian connexions. MACKAY [1965] in his theoretical interpretation has localized in the central encephalon a control system he calls 'normative meta-organization'.

An Experimental Analysis of the Function of Plastic Schemes; Exner's Prepared Reflex

I have worked out since 1964 an electrophysiological model of prepared reflex. The processes thus investigated begin with conscious ideo-representative elaborations. An intentional praxis activity then follows, developing in a complete and detailed fashion. A motor programming is thereby attained, which is novel, since some particular components are created by the subject for the first time. Then, after some repeated executions by the subject, this programming is fixed as a plastic scheme which allows for an automatic performance [cf. PINELLI, 1964].

Disturbances Induced in the Prepared Reflex by Peripheral Troubles Resulting in Irregular and Inconstant Patterns of Afferences

In everyday motor activity the schemes for automatic motility are adapted to the variable parameters of the objects to be handled in the environment and to the gradually moving limbs of the subject himself. This is provided for by exteroceptive and proprioceptive afferences arising from the external contacts and the deplaced tissues. If these consistent patterns of afferences are disturbed, the mechanisms connected with the plastic scheme are upset; in this condition, different forms of 'secondary diskinesias' are observed at the clinical level, such as:

(1) The afferent and efferent innervations which have contributed to the formation of motor schemes are reduced or abolished during compulsory pathological immobility of the limbs. In these conditions the cerebral processes supporting automatic motility tend to disappear. What is called 'Unbefangenheit' [BUMKE, 1948] occurring after long-lasting motor inactivity can be explained according to this theory.

(2) The normal schemes for automatic motility can be blocked by abnormal afferences of the 'algic' type, both with inhibitory effect or over-synchronized excitations. The consequence in the clinical field can be represented by 'secondary functional akinesia'.

(3) The normal motor schemes can no longer have an automatic operation if the connected peripheral patterns are noticeably changed by heterogeneous reduction in the conduction velocity of peripheral nerve fibers (Gombault's neuritis), or by neuritic 'trigger zones' (Kugelberg) or by

variable reversible neuro-apraxic (Seddon) blocks: in reality, paradoxical and misleading messages are then operating.

In all these cases an intentional, ever-newly arising subjective engagement in motor programming is required.

Experiments for the Study of a model on 'Secondary Diskinesias'

In order to make up an experimental model for the reproduction of peripheral abnormal activity with irregular and very instable occurrence I have adopted the ischemia test applied by SIMARD, BASMAJIAN and JANDA [1968]; my purpose was to study in this condition any repercussion, not only on the recruitment of motor units, but also on the prepared reflex.

In 4 normal subjects a prepared reflex [PINELLI, 1964] with a latency of 90 msec was obtained. During ischemia, when the electrical excitation of the nerve above the compressed region gives a repetitive response still of a high amplitude, the prepared reflex previously obtained can no longer be repetitively executed by the subject: the mean latency time of the reactions was 145 msec with a standard deviation of 35 msec.

In some other experiments it was observed that, if ischemia was applied at 15-min intervals during the formation of prepared reflex, the mean latency time of the 'acceleration phase' [PINELLI, 1964] did not drop below 160 msec with a standard deviation of 40 msec.

In both types of experiments with ischemia tests the modality of innervation of antagonist and synergist muscles was markedly disturbed.

The analysis of the results thereby obtained has led to the following interpretation: the repercussions observed in the reactions could not be explained as extinction or external inhibition effects; in reality, abnormal afferent impulses of tetantic type do upset a motor scheme.

A similar series of experiments with the relaxation reactions [PINELLI, 1964] is under way.

Conclusive Remarks

In my opinion this type of research is worthwhile in order to analyze from a neurological point of view some phenomena which are still commonly defined as hysterical insertion, or epithymic superimposition or 'pathologische Reaktionsweise'. In a more remote field of speculation, it may well be that some processes now partaking of psychology of suggestion and ideo-motor aptitude (Eysenck) will be reasonably investigated from a neuro-cybernetical point of view.

References

BUMKE, O.: Lehrbuch der Geisteskrankheiten; 7th ed. (Bergman, München 1948).

CATTELL MACK, J.: Über die Zeit der Erkennung und Benennung von Schriftzeichen, Bildern und Farben. Phylos. St. *2:* 635–650 (1885); Trans. J. MACK CATTELL *1:* 13, 25 (1947).

CATTELL MACK, J.: The time taken up by the cerebral operations. Mind *11:* 220–242, 377–392, 524–538 (1886). — Reprinted J. MACK CATTELL *41:* 94 (1947).

EXNER, S.: Experimentelle Untersuchung der einfachsten psychischen Processe. Pflügers Arch. ges. Physiol. *7:* 600–660 (1873); *8:* 526–537 (1874).

GEORGE, F.H.: Finite automata and the nervous system. Progr. Brain Res., vol. 2, pp. 37–52 (Elsevier, Amsterdam 1963).

GELLHORN, E. and HYDE, J.: Influence of proprioception on map of cortical responses. J. Physiol. *122:* 371–385 (1953).

HEBB, D.O.: The organization of behavior. N.Y. Wiley *110:* 403, 411, 426 (1949).

PINELLI, P.: Risentimenti funzionali sul SNC di lesioni neurologiche periferiche, distrofie riflesse, lesioni viscerali ed algie. Note e Riviste di Psichiatria, Fasc. 2 (April 1959).

PINELLI, P.: Interessi attuali dello studio del 'riflesso preparato' d'Exner-Cattell e della reazione di rilassamento. Riv. Sper. Fren. *88:* 426–450 (1964).

MACKAY, D.M.: Cerebral organization and the conscious control of action. Pont. Acad. Sci., Città del Vaticano, VIII Seminario, 1965, pp. 627–655.

MILNER, P.M.: The cell assembly. II. Psychol. Rev. *64:* 242–252 (1957).

NAYRAC, P.: L'obstacle de non-liéarité en neurologie. Progr. Brain Res., vol. 2, pp. 215–223 (Elsevier, Amsterdam 1963).

PENFIELD, W. and JASPER, H.H.: Epilepsy and the functional anatomy of the human brain (Churchill, London 1954).

PHILLIPS, C.G.: Changing concepts of the precentral motor area. Pont. Acad. Sci., Città del Vaticano, VIII Seminario, 1965, pp. 579–626.

PURVES-STEWART, J.: The diagnosis of nervous diseases; 9th ed., pp. 443–446 (Arnold, London 1947).

RUSSEL BRAIN: La notion de schème en neurologie et en psychiatry; in Perspectives cybernétiques en psychophysiologie (Presse universitaire de France, Paris 1951).

SIMARD, T.G.; BASMAJIAN, J.V. and JANDA, V.: Effects of ischemia on trained motor units. Amer. J. phys. Med. *47:* 64–71 (1968).

VON NEUMAN: The general and logical theory of automata; in JEFFRESS Cerebral mechanism in behavior. Hixon Symp. (Wiley, New York 1951).

WOODWORTH, R.S. and SCHLOSBERG, H.: Experimental psychology, pp. 541–581 (Rinehart & Winston, New York 1968).

WUNDT, W.: Grundzüge der physiologischen Psychologie; 2nd ed. (1880).

Author's address: Dr. P. PINELLI, Università Cattolica Clinica Neurologia, Via Peneta Sachetti 526, *Rome 00168* (Italy)

Medicine and Sport, vol. 6: Biomechanics II, pp. 39–47
(Karger, Basel 1971)

Experiences with Spinal Reflexes in Research of the Human Motor System

C.H.M. Brunia

De Klokkenberg, Breda

Introduction

Whereas electromyography is a very valuable tool in clinical neurology, its use in psychophysiological research seems limited. Evoking spinal reflexes has the advantage that one is stimulating a comparatively well-known structure in the nervous system. This knowledge is based upon experiments with anaesthetized animals. Although one does not know what connection this has with normal animals, let alone human beings, reasonable hypotheses about what is going on in the nervous system are possible. The work we have done up to now is limited to the gastrocnemius and soleus muscles.

By an electric stimulation of the 1A afferent fibres of these muscles, their α-motoneurons are activated. The afferent volley causes a contraction of the muscle, the potential shift of which can be registered. The amplitude of the triphasic wave is supposed to be a measure of the activity of the α-motoneuronpool (Hoffmann reflex [HR]).

If a tap is given on the tendon, the muscle spindles are lengthened and the receptors in the muscle spindle are stimulated. Here also the α-motoneurons of the muscle are activated by means of the 1A afferent fibres. The sensitivity of the receptors is regulated by the γ-motoneurons, so that the amplitude of the T-reflex is an expression of both α- and γ-activity (Achilles tendon reflex [TR]).

The Simultaneously Evoked HR and TR

In case both reflexes are evoked simultaneously it is possible to compare the possible different way by which both kinds of motoneurons are influenced. Such systematic psychophysiological research has been initiated by Paillard [1955].

Requin [1967, 1969] used the technique to explore the variations of excitability of the motoneuron-pools during the pre-period of a reaction-time experiment. We did different experiments to evaluate the influence of alertness on the spinal level by giving subjects a continuous performance task [BRUNIA, 1969]. One of them will be discussed here.

Methods and Material

Stimulation and Recording

The subject is lying on a bed in a prone position. Mechanical stimulation is done by a hammer, placed some millimeters above the Achilles tendon. Electrical stimulation consists of rectangular pulses of 1 msec and an individual strength, depending among others on the skin resistance. Stimuli are delivered to the skin over the popliteal fossa. Recording is done by surface electrodes over the gastrocnemius muscle in both limbs. The reflexes are visualized by a double beam oscilloscope of which the time base has been stopped. They appear as vertical lines and are recorded on photographic paper (fig. 1).

Task

The binary choice generator, an apparatus developed in the Institute of Perception Research at Eindhoven [VALBRACHT, 1960] was used for the performance task. In a random succession one of two light-bulbs illuminates. The subject has to press a button he has in his right or left hand, depending on whether the light illuminates in his right or left field of vision. The frequency with which the bulbs illuminate is variable from person to person but we always chose one with which it is too difficult to give only correct answers. Otherwise the task is not so difficult that it becomes impossible to give correct answers. Having fixed the frequency for a subject, we keep it constant during the experimental period of 20 min. Reflexes are evoked at a frequency of 1 per 3 sec.

Statistical Procedure

Mean amplitudes and standard deviations are calculated for periods of 5 min. The results are transformed into Z scores for every subject. The Z scores are added up and the mean Z score is computed for every period of 5 min. The 0.95-level of confidence is represented by the vertical lines in figure 2.

Experiment

In this particular experiment we were not interested in the relation, if any, between the subject's performance and the amplitude of his reflexes.

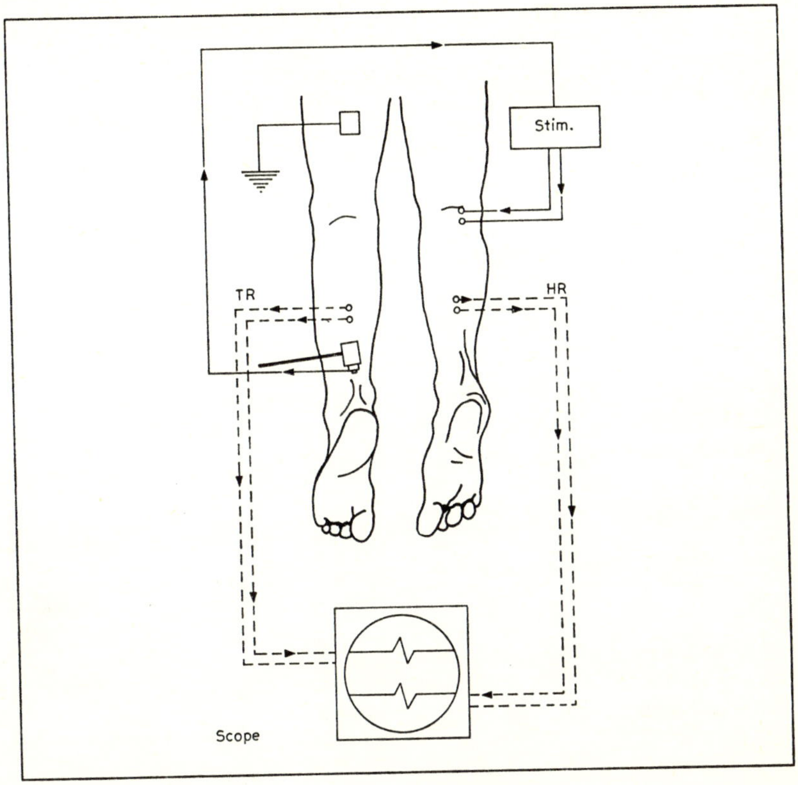

Fig. 1. The experimental situation. The hammer falls on one of the Achilles tendons. At the very contact with the skin electric stimulation at the popliteal fossa of the other leg is started by means of a microswitch in the head of the hammer.

The task was used to create a situation in which alertness was necessary to produce correct answers. We wanted to know what kind of influence the alertness had on the TR and HR, so we had to compare the amplitudes at rest and during alertness.

The total length of the experiment was 40 min. After a 10-min rest period, the task was presented to 19 healthy subjects for 20 min. The experiment was terminated by a second period of 10 min rest.

Results

The results are summarized in figure 2.

1. TR. The mean amplitude during the task shows a highly significant increase ($p < 0.001$).

2. HR. There is no significant difference between the amplitudes at rest and during the task ($p > 0.20$).

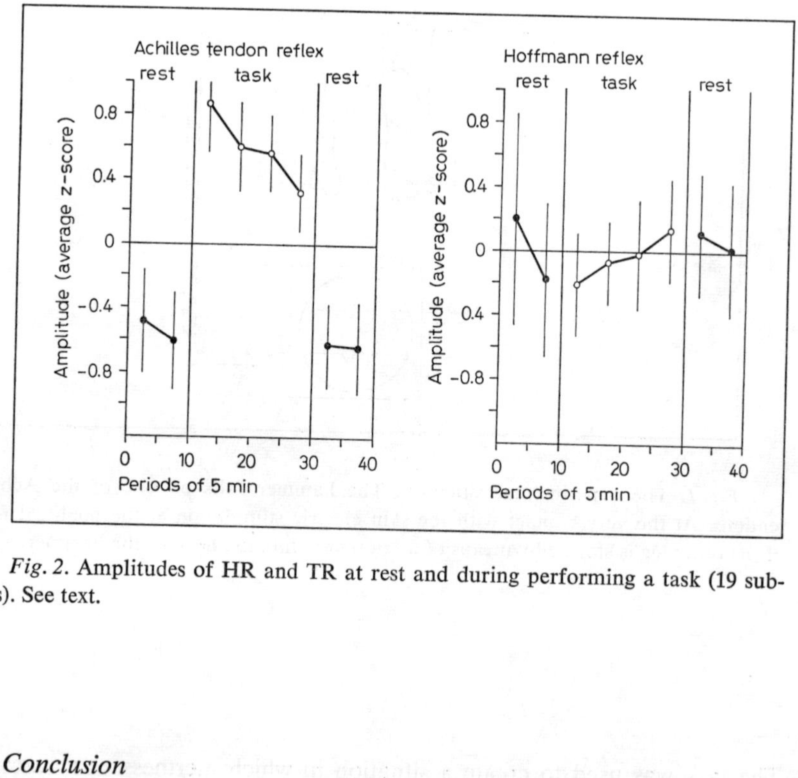

Fig. 2. Amplitudes of HR and TR at rest and during performing a task (19 subjects). See text.

Conclusion

The higher amplitudes of the TR during the task without clear changes of the HR seem to be explained by a hyperactivation of the γ-motoneurones. We are in fact confronted with a hyperactivation of the dynamic fusimotor system during the task, because especially in the mechanically evoked reflexes the dynamic fusimotor system is tested. Probably this is due to a generalized nonspecific effect of reticular formation activity [BRUNIA, 1971].

The Recovery Cycle of HR and TR

In case two electric stimuli (S_1 and S_2) are given to the afferent fibres of the tibial nerve in the popliteal fossa, two HR (H_1 and H_2) are to be recorded. The amplitude of H_2 depends among others on the length of the interval between S_1 and S_2. The amplitude of H_2 may be expressed as a percentage of the amplitude of H_1. The relation between the stimulus interval and the amplitude of H_2 is shown in the recovery cycle (fig. 3). The amplitude of H_2 does not simply increase with the enlargement of the interval between S_1 and S_2. After a total supression for about 75 msec ('early depression') facilitation occurs between 100 and 225 msec. A second ('late') depression follows from 225 to 300 msec, after which a slow facilitation sets in. Complete recovery is not yet attained at 1,000 msec.

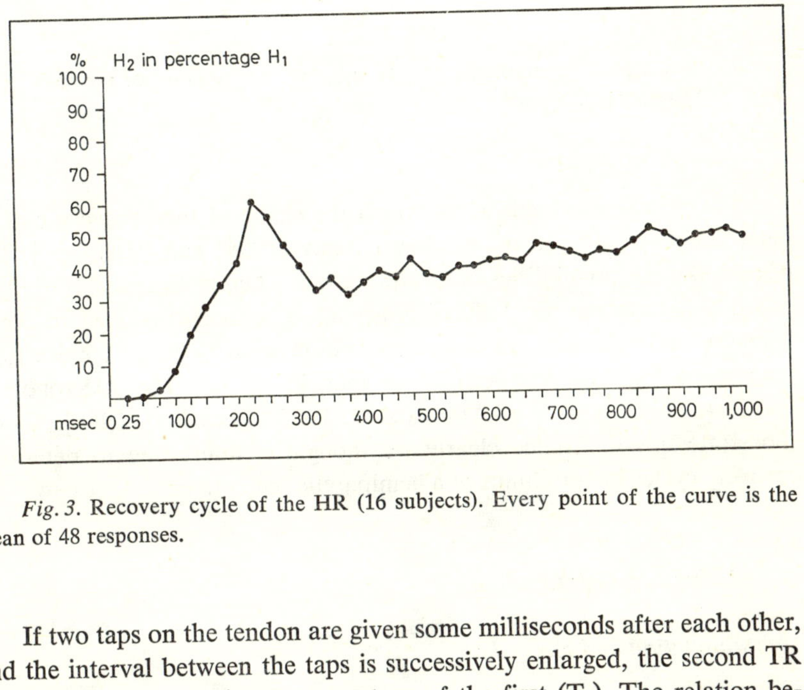

Fig. 3. Recovery cycle of the HR (16 subjects). Every point of the curve is the mean of 48 responses.

If two taps on the tendon are given some milliseconds after each other, and the interval between the taps is successively enlarged, the second TR (T_2) may be expressed as a percentage of the first (T_1). The relation between stimulus interval and the amplitude of T_2 is shown in the recovery cycle (fig. 4). Recovery of the TR does not take place at a stimulus interval less than 250 msec, thus not before the late depression of the HR has started. Moreover, it passes faster and reaches a higher level.

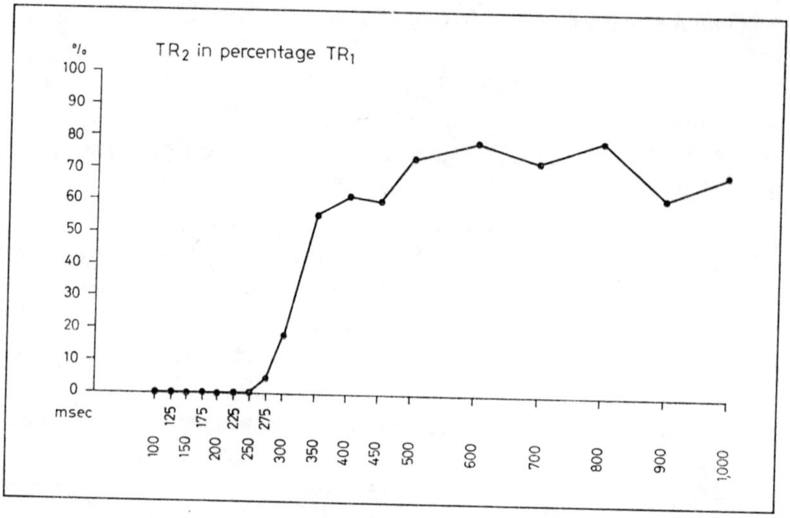

Fig. 4. Recovery cycle of the TR (18 subjects). Every point of the curve is the mean of 54 responses.

The precise mechanisms by which the shape of the recovery cycle is determined are not yet known. MAGLADERY [1954] and PAILLARD [1955] discussed some possibilities of interpretation, whereas ECCLES [1967] proposed a hypothesis about the supraspinal influences, involved in the H recovery cycle. In patients with uppermotoneuron diseases a more rapid recovery of the HR has been found than in normals [DIAMANTOPOULOS and ZANDER OLSEN, 1966; MAGLADERY, 1952; TAKAMORI, 1967]. To demonstrate this difference clearly we thought it interesting to determine recovery cycles in both limbs of a hemiparetic patient.

Methods and Material

Stimulation and Recording

The technique used in stimulation and recording responses is *mutatis mutandis* the same as described above.

Three pairs of electric stimuli are given at every interval S_1–S_2. The intervals vary from 100 to 1,000 msec in steps of 100 msec. Only from 100 to 200 msec we use steps of 25 msec. Time between S_1 and the following S_1 is 15 sec. Recovery cycles in both limbs are determined in 6 sessions on 6 different days. H_2 is expressed as a percentage of H_1. The mean percentage and standard errors are calculated.

The same procedure is followed with mechanical stimulation. Here we use steps of 25 msec from 100 to 300 msec. T_2 is expressed as a percentage of T_1. Here too, mean percentages and standard errors are calculated.

Experiment

Of an 18-year-old right-sided hemiparetic patient recovery cycles of HR and TR have been determined in both legs, each in six different sessions.

Results

The results are demonstrated in figures 5 and 6. It appears that a higher facilitation level is reached in the hemiparetic limb. This concerns both H and T reflexes.

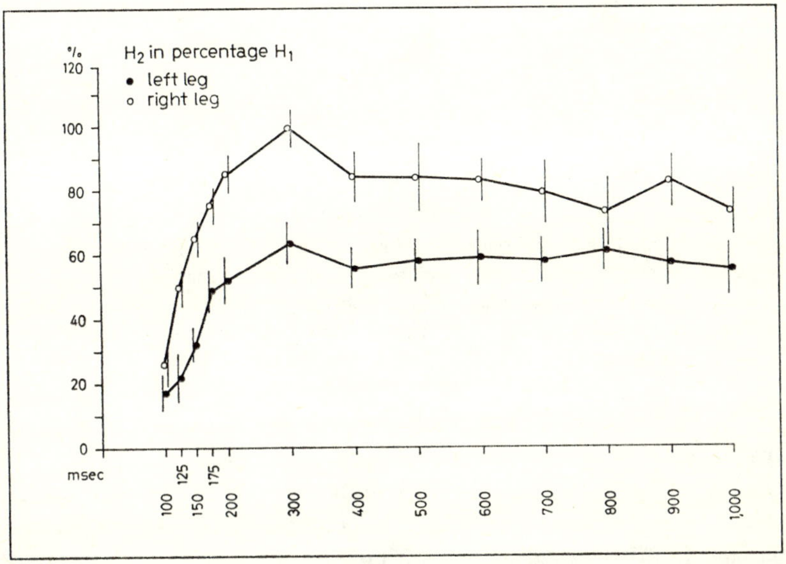

Fig. 5. Recovery cycle of the HR of a patient with a right-sided hemiparesis. Every point of the curve is the mean of 18 responses.

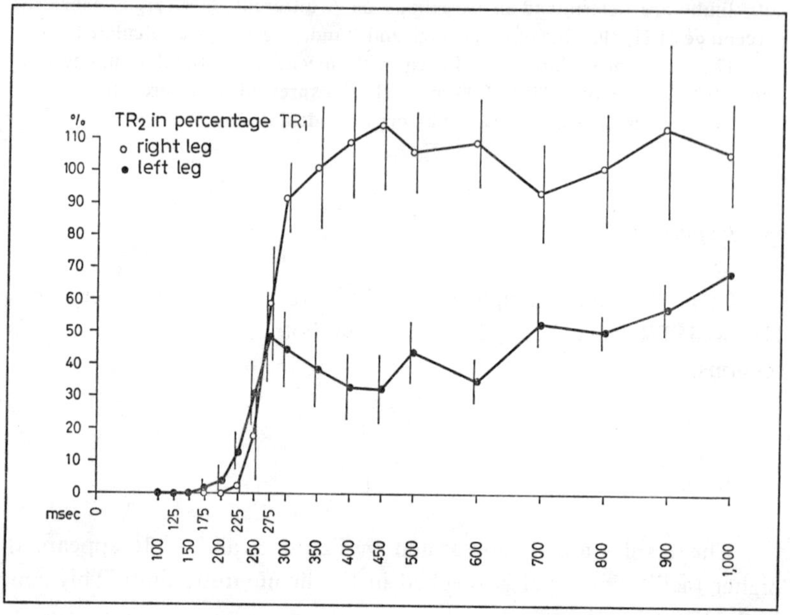

Fig. 6. Recovery cycle of the TR of a patient with a right-sided hemiparesis. Every point of the curve is the mean of 18 responses.

Conclusion

Because facilitation is seen in both kinds of recovery cycles, the conclusion seems warranted that the recorded effect in the hemiparetic limb affects the α-motoneuron system. The central and peripheral factors, that determine the configuration of the curves, are presumably not influenced by the disorders of the upper motoneuron system in this patient. The α-motoneuron complex of the affected side might be influenced directly by facilitation from the higher levels.

From the preliminary results, mentioned by VREDENBREGT at the symposium it might be concluded that using his stimulation apparatus hemiparetic patients show a real improvement. If this is true, something must have changed in the nervous system of these patients during the period the apparatus was used. Perhaps it may be possible in the future to show by means of the methods here mentioned how these improvements have been realized.

Acknowledgement

It is a pleasure to thank TON VAN BOXTEL and WIM KNYFF for their assistance during the experiments.

References

BRUNIA, C.H.M.: The influence of a task on the amplitude of the Achilles tendon reflex and the Hoffmann reflex. Acta physiol. pharmacol. neerl. *15:* 395–396 (1969).

BRUNIA, C.H.M.: To be published in Physiology and Behavior. 1971.

DIAMANTOPOULOS, E. and ZANDER OLSEN, P.: Motoneuron excitability in patients with abnormal reflex activity; in GRANIT Muscular afferents and motor control. Nobel Symposium I, pp. 451–452 (Almqvist & Wiksell, Stockholm 1966).

ECCLES, J.C.: The way in which the cerebellum processes sensory information from muscle; in YAHR and PURPURA Neurophysiological basis of normal and abnormal motor activities, pp. 379–406 (Raven Press, New York 1967).

MAGLADERY, J.W.; TEASDALL, R.D.; PARK, A.M., and LANGHUTH, H.W.: Electrophysiological studies of reflex activity in patients with lesions of the nervous system. I. Bull. Johns Hopk. Hosp. 219–244 (1952).

PAILLARD, J.: Reflexes et régulations d'origine proprioceptive chez l'homme (Arnette, Paris 1955).

REQUIN, J.: Organisation de l'activité motrice préparatoire; in DESSART Hommage à ANDRÉ REY, pp. 337–366 (Soledi, Liège 1967). – Some data on the neurophysiological processes involved in the preparatory motor activity to reaction time performance; in KOSTER Attention and performance, vol. 2, pp. 358–367 (North Holland Publishing Comp., Amsterdam/London 1969).

TAKAMORI, M.: H-reflex study in upper motoneuron diseases. Neurology *17:* 32–40 (1967).

VALBRACHT, J.: IPO ontwerpspecificatie Nr. 9 (Instituut voor Perceptie Onderzoek, Eindhoven 1960).

VREDENBREGT, J.: A functional muscle stimulator for hemiplegic patients. Medicine and Sport vol. 6, pp. 285–287 (Karger, Basel 1971).

Author's address: Dr. C. H. M. BRUNIA, Dr. HANS BERGER Kliniek, De Klokkenberg, *Breda* (Netherlands)

Medicine and Sport, vol. 6: Biomechanics II, pp. 48–54
(Karger, Basel 1971)

Fusimotor Neurons in Voluntary Movement[1]

T. W. ROBERTS, J. L. SMITH and E. MORTIMER ROBERTS
University of Wisconsin, Madison, Wisc.

Extreme dynamic range and the high resolution of human fast voluntary movement suggest a special temporal interplay of higher neural processes and the physio-mechanical properties of skeletal muscle.

The role of the cerebral cortex in the control of intricate movement in monkey and man is well documented. There is a corticospinal monosynaptic pathway to the α-motor neuron, presumably for direct cortical control of voluntary movement. In addition, GRANIT and KAADA [1952] demonstrated that the fusimotor or γ-motor neuron receives a large contribution of higher central control. That the muscle spindle is a powerful regulator of muscle length has been demonstrated by LIDDELL and SHERRINGTON [1924], GRANIT [1968], and others. Models of voluntary regulation of skeletal muscle known as 'servo'-control or, more broadly, α-γ coactivation, have been built around the muscle spindle and the stretch reflex.

That individual body-segment participation in voluntary movement falls in the range of tens of milliseconds has been repeatedly shown by GLASSOW [1968], ROBERTS [1968], ATWATER, and others at Wisconsin through detailed cinematographic analysis of human athletic skills. For example, although the total sequence of movements in the overarm throw pattern may require over a second in time, all contributions to forward motion occur within 300 msec and the convergences of forces accelerating the ball are delayed until the last 40–70 msec. ATWATER [1969] has found that peak angular velocities of 6,000 °/sec and accelerations of 300,000 °/sec^2 are characteristic of elbow joint action in the skilled male performer. Since

1 Supported in part by NIH grant No. NB-04864 and in part by the Research Committee of the University of Wisconsin.

it is known that the fusimotor feedback loop requires 20–50 msec, these data raised fundamental questions about the capacity of the fusimotor system as a feedback element participating in fine motor control during rapid movement. On the other hand, the presence of large numbers of muscle spindles in the arm and hand, the very focus of rapid movement, invited further consideration.

To elucidate fusimotor participation in voluntary movements of the hand and arm, (1) a study of the degree of cortical regulation of γ-motor neurons in cervical spinal ventral roots of cats and monkeys, and (2) an examination of the effects of anesthetic blockade of fusimotor neurons in the radial nerve on performance of human motor tasks were conducted.

In the animal series 25 cats and 3 monkeys were anesthetized with sodium pentabarbital or chloralose-urethane intraperitoneally. An anterior cordotomy exposed one or two of cervical ventral roots C-5 through C-8. A craniotomy was performed over the contralateral pericruciate or pericentral cerebral cortex. Dual matched tungsten micro-electrodes carried in a stereotactic instrument were used to locate single fusimotor fibers in the ventral roots. Low level diphasic electrical stimulation at 100 pulses/sec was applied to cortical points through a monopolar electrode.

In the cervical roots, fusimotor neurons were found to fire at 10–15 pulses/sec. This range is more restricted than the firing rates 0–80 pulses/sec found in lumbar roots by HUNT [1951] and by MORTIMER and AKERT [1961]. In contrast, α-motor neuron firing was rarely seen in the brachial plexus, except intermittently during spontaneous movement or prolonged low level stimulation of the cortex. Fusimotor neurons were verified by their tendency toward continuous firing, and their small potential size, and when possible, by antidromic conduction velocities.

Peripheral stimulation such as touch, pressure to the limb or footpad or joint movement had a limited and very transitory effect on fusimotor firing rates; responses dropped out 100–300 msec after cessation of stimulation.

Electrical stimulation of the motor cortex yielded maps of cortical activation of fusimotor fibres similar to the map for cortically evoked limb movement by WOOLSEY [1958] and others (fig. 1). In sharp contrast to fusimotor fibres in lumosacral roots to the hind leg, the most striking result of cortical stimulation of cervical fusimotor neurons was the immediacy of response (frequently less than 40 msec) to onset and cessation of cortical stimulation. MORTIMER and AKERT [1961] found that many of the lumbar fibers required half a second or more to reach peak firing and long after discharges often persisted upon cessation of cortical stimulation.

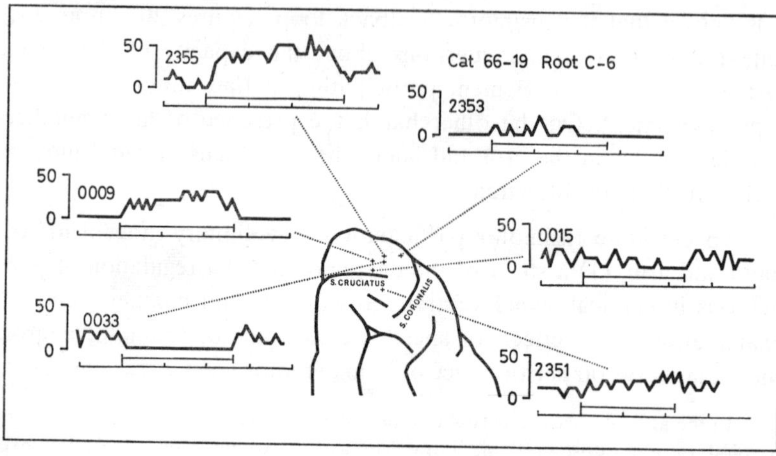

Fig. 1. Map of responses of C-6 fusimotor fiber to cortical stimulation. Cat under chloralose-urethane.

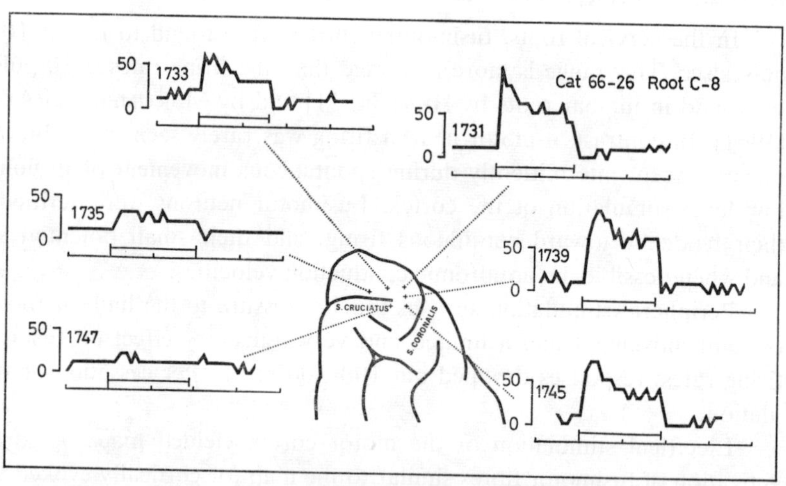

Fig. 2. Map of responses of C-8 fusimotor fiber to cortical stimulation. Animal curarized by Flaxedil. Cat under chloralose-urethane.

Another interesting piece of cortical control data was the limit of firing rates of cervical fusimotor neurons. Although cortical stimulation yielded sharp responses, steady firing beyond 40–50 pulses/sec was rarely seen. That an inhibitory mechanism may be working here is indicated in figure 2. In this experiment the cat was under Flaxedil paralyzing body musculature. Under these conditions the cortical map for activation of the unit became very large and no inhibitory responses could be found. Cortical activation reached peaks of 100 pulses/sec.

The idea we want to suggest here is that cervical fusimotor neurons are easily activated by the cortex and under normal conditions, once activated, become the source of their own limitation in firing rate so that they do not dominate the control of movement in servo-fashion. Rather, this inhibitory process assists in converging fusimotor facilitation on selected motor units during voluntary movement. We sought to test these ideas by examining the effects of anesthetic blockade of fusimotor neurons on performance of human motor tasks.

Behavioral studies were conducted to compare the motor behavior of normal females performing voluntary movement tasks with intact and blocked fusimotor fibers of the right radial nerve [SHAMBES, 1969].

A differential radial nerve block was obtained by infiltrating 15 ml of 0.4% xylocaine solution approximately 5 cm from the axillery apex [BONICA, 1953]. The local anesthetic differentially affected the small nerve fibers, A-γ through C [MATTHEWS and RUSHWORTH, 1957]. An A-γ block, in the presence of viable A-α nerve fibers, was inferred [LANDAU *et al.*, 1963] by (1) absence of triceps tendon reflex, (2) loss of temperature sensations and reduction of pain sensations (light pin prick) at radial cutaneous areas, and (3) normal strength responses of the elbow and wrist extensors (fig. 3). Subjects were not tested on the motor tasks until these criteria were satisfied.

Motor tasks involved the affected musculature (elbow extensors) as agonists or antagonists and had differential requirements for speed and accuracy. The tasks included: (1) reaction time of elbow extension and flexion, (2) dart throwing, (3) finger-to-nose placement test, and (4) reciprocal movements at 2 different speeds (0.33 and 1.6 Hz).

Dysmetric motor behavior was typically observed when subjects attempted to perform the finger-to-nose placement test at their normal speed. Despite the opportunity for 'visual control', the nose was missed, and the mouth or face was hit. Contacts were forceful and unexpected as evidenced by the subjects' facial expressions. These results indicate that elbow flexion was not damped sufficiently when the fusimotor innervation of the elbow extensors was blocked. In addition, performance on the reaction time and dart throwing tasks indicated that the contribution made by the fusimotor neurons in initiating rapid movement was lost. The average increase in reaction time for elbow extension was 20 msec. The first dart throws

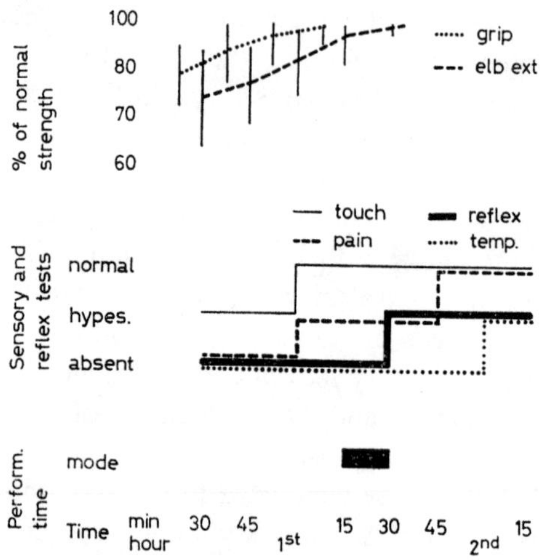

Fig. 3. Recovery sequence of the radial nerve following xylocaine infiltration. Median strength scores, modal sensory responses and triceps tendon reflex indicate that a differential block occurred between 60 and 90 min for most subjects.

averaged 0.75 m/sec slower than normal and landed 12 cm lower on the target. Films of the subjects performing the dart throw revealed that the reduction in initial velocity was due to a decrease in the acceleration of elbow extension. During later throws, increased shoulder action appeared to compensate for the subjects' inability to rapidly accelerate the forearm (fig. 4).

In conclusion, it is our feeling that the fusimotor system does not behave in a servo-fashion during fast voluntary movement. The power of the corticospinal pathway to the α-motor neuron is great. The fusimotor system is fast in its cortical responses but owing to its limitation in peak firing cannot contribute full servo driving to the α-motor neuron via afferent fibres. What is more likely is that the fusimotor system is involved in the process of 'commutation', that is in selecting the exact moment that a subliminal motor unit shall be activated and the exact moment it shall cease contributing to voluntary movement in response to corticospinal influences. In behaving in this way the fusimotor system may be just as important where it is silent, as its lack of facilitation of a non-participating motor

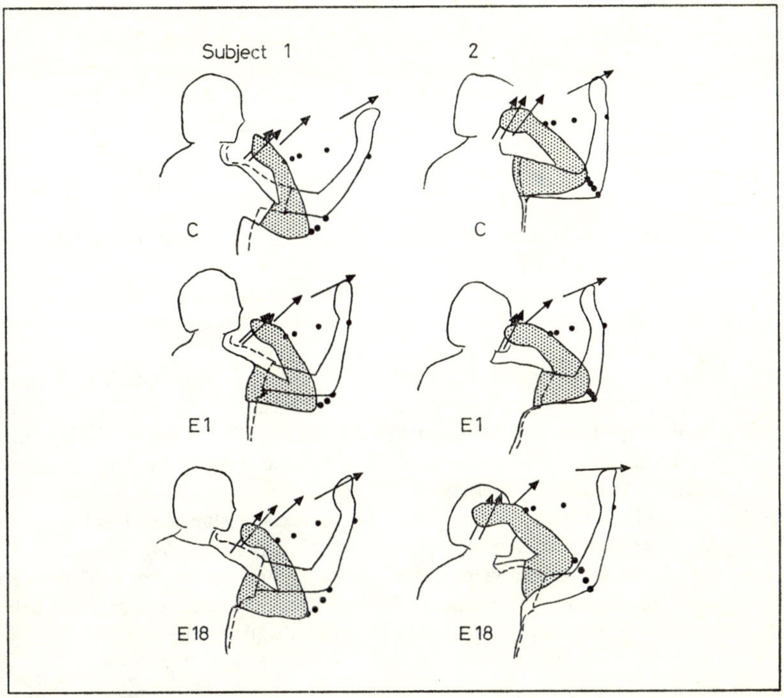

Fig. 4. To throw the dart the subject stood right side facing a target placed 2.5 m away. Composite tracings from 16-mm films of two subjects taken before (C) and after (E) the differential radial nerve block illustrate the initial decrease in forearm displacement (E 1) and the increased shoulder action 18 throws later. Dots show the position of the wrist and elbow at 45-msec intervals prior to dart release.

unit would increase the final common path-safety factor, making sure it does not fire in response to local proprioceptive influences. During the middle of activation of a motor unit the fusimotor system may contribute or subtract nothing, but may be most important in regulating the beginning and end of segmental contribution to voluntary movement.

References

Atwater, Ann E.: Personal communication (1969).
Bonica, J. J.: The management of pain (Lea & Febiger, Philadelphia 1953).
Cooper, J. M. and Glassow, R. B.: Kinesiology; 2nd ed. (Mosby, St. Louis 1968).

GRANIT, R.: The functional role of the muscle spindle's primary end organs. Proc. roy. Soc. Med. *61:* 69–78 (1968).

GRANIT, R. and KAADA, B. R.: Influences of stimulation of central nervous structures on muscle spindles in cat. Acta physiol. scand. *27:* 130–160 (1952).

HUNT, C. C.: The reflex activity of mammalian small-nerve fibers. J. Physiol., Lond. *115:* 456–469 (1951).

LANDAU, W. M.; WEAVER, R. A., and HORNBEIN, T. F.: Fusimotor nerve function in man. Arch. Neurol., Chicago *3:* 10–23 (1963).

LIDDELL, E. G. T. and SHERRINGTON, C. S.: Reflexes in response to stretch (myotatic reflexes). Proc. roy. Soc. B *96:* 212–242 (1924).

MATTHEWS, P. B. C. and RUSHWORTH, G.: The selective effect of procaine on the stretch reflex and tendon jerk of soleus muscle when applied to its nerve. J. Physiol., Lond. *135:* 245–262 (1957).

MORTIMER, E. M. and AKERT, K.: Cortical control and representation of fusimotor neurons. Amer. J. phys. Med. *40:* 228–248 (1961).

ROBERTS, E. MORTIMER and METCALFE, A.: Mechanical analysis of kicking; in WARTENWEILER, JOKL and HEBBELINK Biomechanics: Medicine and Sport, vol. 2, pp. 315–319 (Karger, Basel 1968).

SHAMBES, G. M.: Influence of the fusimotor system on stance and volitional movement in normal man. Amer. J. phys. Med. *48:* 225–236 (1969).

WOOLSEY, C. N.: Organization of somatic sensory and motor areas of the cerebral cortex; in HARLOW and WOOLSEY Biological and biochemical bases of behavior, pp. 63–81 (University of Wisconsin Press, Madison 1958).

Authors' addresses: THOMAS W. ROBERTS and ELIZABETH MORTIMER ROBERTS, The University of Wisconsin, Department of Physical Education for Women, *Madison, WI 53706;* JUDITH L. SMITH, Ph.D., The University of California, Department of Physical Education, *Los Angeles, CA 90024* (USA)

Medicine and Sport, vol. 6: Biomechanics II, pp. 55–59
(Karger, Basel 1971)

The Muscle Spindle as a Control Element

K. L. Boon

Twente University of Technology, Enschede

Introduction

In describing the behaviour of a muscle as a second-order process with a time-delay, control engineering shows how such a process can best be regulated. Generally, the muscle spindle is considered as a regulator of muscle action. In this article muscle spindle is described from this control engineering point of view.

In figure 1a general model of a part of the neuromuscular system is given. It shows a control system consisting of a controlled process (the muscle) and a controller (the muscle spindle). Besides being a controller, the muscle spindle is also a transducer. The figure also shows (1) the α-motor neurons which activate the muscle, (2) the γ_s- and γ_d-efferents, (3) I_a- and II-afferents, (4) spinal cord.

The aim of this article is to review the facts known from the literature about muscles and muscle spindles and to fit these facts in a simple general concept.

a) A Model of a Muscle

Many authors have described the behaviour of a muscle by means of a model containing a force generator, spring, damper and inertia. Although this model does not cover all experimental facts, it is often used for calculating purposes [Vickers, 1968; McRuer *et al.*, 1968]. The spring and damper both depend on the active state of the muscle. In figure 2 an example of such a model is given.

The model can be described by means of a second-order process. It is a process that is generally determined by two characteristic time

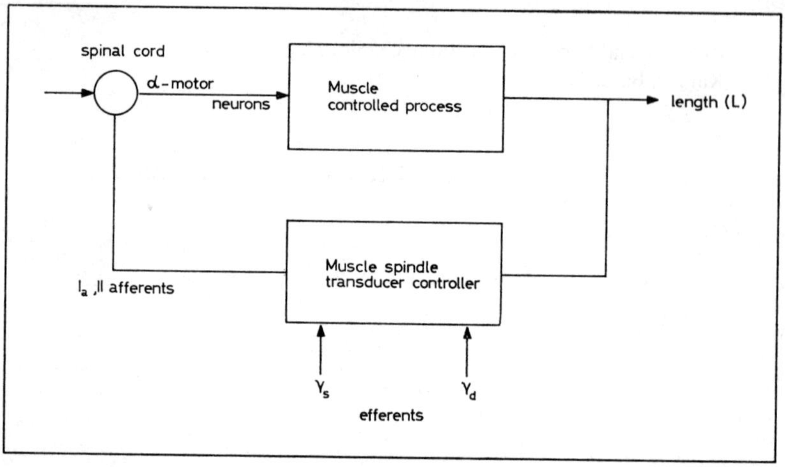

Fig. 1. A general model of a part of the neuromuscular system.

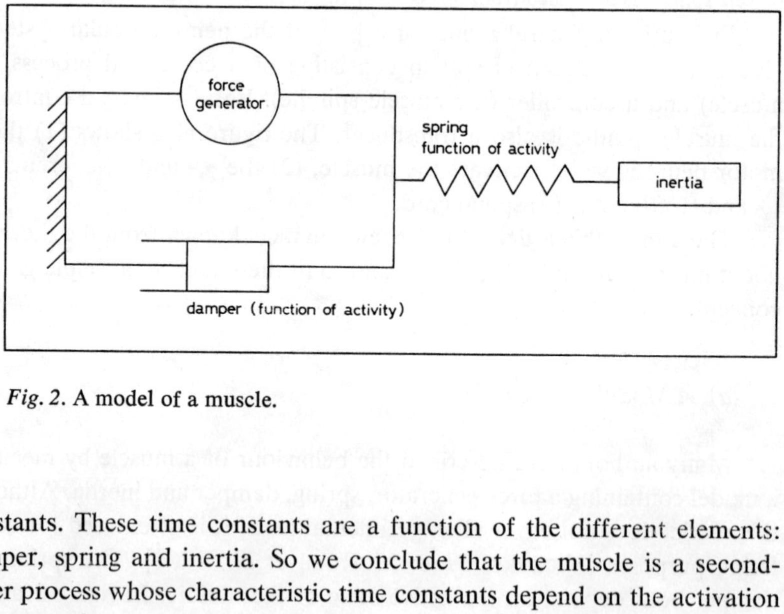

Fig. 2. A model of a muscle.

constants. These time constants are a function of the different elements: damper, spring and inertia. So we conclude that the muscle is a second-order process whose characteristic time constants depend on the activation of the muscle.

When we consider the muscle *in vivo* as part of the general system (fig. 1) a time delay has to be taken into account.

b) A Model of a Muscle Spindle

The rather complicated structure of the muscle spindle is apparently the cause of many disagreements about its morphology and anatomy as well as its physiology.

Figure 3 shows a part of figure 1 in which the muscle spindle represents a system with three inputs: length, static γ-efferents (γ_s) and dynamic γ-efferents (γ_d), and two outputs: the I_a- and the II-afferents. The names γ, I_a and II are indications used in muscle spindle literature.

The names static and dynamic γ-fibres refer to properties which can be observed by means of certain experiments.

We will consider only the I_a-output as being most important. Its activity depends on the variables length (L), γ_s and γ_d. Many experiments [CROWE and MATTHEWS, 1961; MATTHEWS, 1964] have established that the activity of I_a is proportional to the length (engineers call this activation proportional control), and also that the activity depends on the velocity (dL/dt) (we call this activation derivative control).

The sensitivity to length and velocity can be varied by changing the γ_s- and γ_d-activities.

We assume that there must also be an integral control (this is important for slow changes of length e.g. in posture control). This means that the I_a-activity depends on the integral $\int$ Ldt too. We conclude that the muscle spindle is a control element whose output can be described mathematically by the formula:

$$I_a = I_a \{\int^{\tau} L(t)dt, L(t), dL(t)/dt, \gamma_s, \gamma_d\}.$$

In this formula the factor $\int^{\tau}$ L(t)dt describes an integral control function (I-control). The value L(t) indicates a proportional control function (P-control) and the factor dL(t)/dt shows that the I_a-activity can depend on the velocity dL/dt which is the derivative control function (D-control). The control actions can be varied by changing the γ_s- and/or γ_d-activities.

c) Synthesis

The synthesis of section a and b is: the properties of the muscle and the properties of the muscle spindle fit well together considering the muscle spindle as a control element of the muscle action.

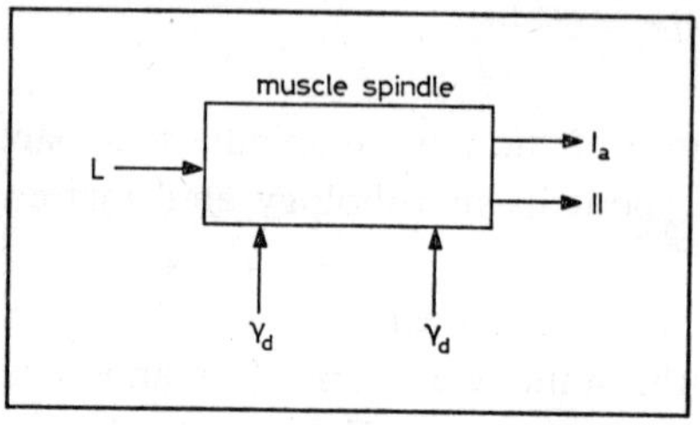

Fig. 3. A model of a muscle spindle.

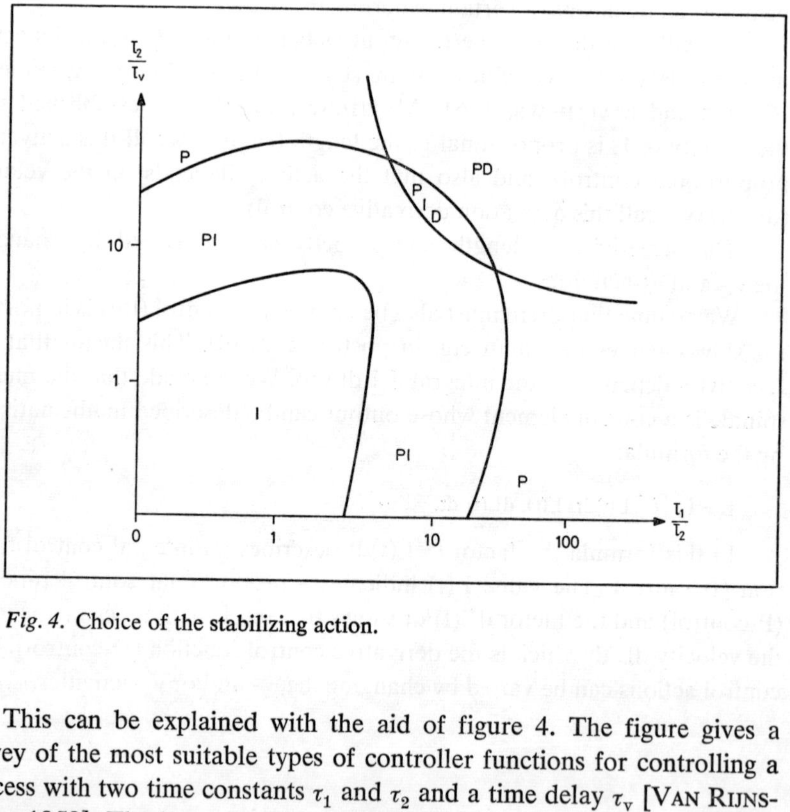

Fig. 4. Choice of the stabilizing action.

This can be explained with the aid of figure 4. The figure gives a survey of the most suitable types of controller functions for controlling a process with two time constants τ_1 and τ_2 and a time delay τ_v [VAN RIJNSDORP, 1959]. The letters P, D and I indicate proportional, derivative and integral control.

In section a it is shown that the time constants depend on the activation of the muscle, so it must be possible to change the control action for good regulation.

In section b we saw that in the muscle spindle this control action can be selected by means of the γ_s- and γ_d-activities. It seems possible to calculate when there will be a change in the γ_s- and γ_d-activities, if we know τ_1, τ_2 and τ_v as a function of the muscular activation. Whether these changes actually occur as expected should be a subject of further investigation. Research of the muscle spindle with this aim in mind looks rather fruitful, because it would make possible a better understanding of the role of the muscle spindle.

References

CROWE, A. and MATTHEWS, P.B.C.: The effects of stimulation on static and dynamic fusimotor fibres on the response to stretching of the primary endings of muscle spindles. J. Physiol., Lond. *174:* 109–131 (1961).

MCRUER, D.T.; MAGDALENO, R.E., and MOORE, G.P.: A neuromuscular actuation system model. IEEE Trans. Man-Machine Systems *9:* 61–71 (1968).

MATTHEWS, P.B.C.: Muscle spindles and their motor control. Physiol. Rev. *44:* 219–288 (1964).

RIJNSDORP, J.E.: in GRABBE and RAMO Handbook of automation, computation and control, vol. 3 (Wiley, New York/London 1959).

VICKERS, W.M.: A physiologically based model of neuro-muscular dynamics. IEEE Trans. Man-Machine Systems *9:* 21–23 (1968).

Author's address: K. L. BOON, Twente University of Technology, *Enschede* (Netherlands)

II. Muscular Co-ordination

Medicine and Sport, vol. 6: Biomechanics II, pp. 62–71
(Karger, Basel 1971)

Control System Aspects of Muscular Co-ordination

J. H. MILSUM

BioMedical Engineering Unit, McGill University, Montreal, Quebec

I. Introduction

Our human neuromuscular systems for limb movements are characterized by much richness of structure so that the key question in neuromuscular co-ordination is not 'Can one find a set of muscles which will do this job when suitably programmed?', but rather 'Which organizing principle or optimizing strategy should one use to select a set of muscles to do the task, in the sense that this is best in some reasonably definable quantitative sense'? Certainly, it is clear to us that many substitution patterns are available by which desired movements can be achieved, when some part of the normally-used muscle system is inoperative through injury, etc.; it is also subjectively obvious to us that such substitution patterns typically would not be optimal for the uninjured case. In this introductory review we shall not consider the detailed questions of how particular sets of muscles are coordinated neurophysiologically, but rather how the overall co-ordinating operation may be programmed temporally, both as to the initial learning and to online adaptation. Such a review must necessarily be largely speculative, since appropriate physiological data are generally lacking.

II. Evolutionary Aspects

In the evolutionary process, selective pressure operates so as to benefit the more fit mechanism, and furthermore the neuromuscular systems provide the key mechanism for the survival of higher animals, whether of the predator obtaining his food, or the prey escaping. Consequently we can expect to be able to infer some reasonably plausible criteria for the performance of the neuromuscular system, and perhaps also for the evolutionary development of a particular system. In this we presuppose that selective pressure has benefited those organisms whose neuromuscular systems provide considerable flexibility in either predation or escape operations. The first basic fact which we must note is that the acquiring of energy through feeding requires both a 'capital investment' and an 'operating cost' for the organism. The capital cost arises in devoting part of the organism's limited resources during growth to the generation of a particular organ system such as the neuromuscular system. The operating cost arises in sustaining and replacing, among other things, the living cells of these systems during the organism's lifetime. Hence it seems clear that when an organism can perform its food-procuring task with less expense of energy than other similar organisms, then, other factors being equal, such organisms will have selective advantage within their species, and furthermore, such species will have in general selective advantage as compared with other less efficient species.

In order to be more specific, we note that in neuromuscular systems producing locomotion, there are at least two, and sometimes more, 'free' actuating variables which must be set by the system's control; for example, the pace length and frequency for walking at a given speed. It seems reasonable to suppose that some particular pair of these parameter values, or perhaps some small operating region, will require less energy than the others, and, therefore, would be selected for in the long process of evolution. In detail, for walking, a quadratic function can specify the oxygen cost to the organism quite well [MILSUM, 1966]:

$$P_V = a + bV^2,$$

where P_V = the rate of oxygen consumption attributable to the locomotion, V = speed, and a and b are constants.

This can be expressed in terms of the oxygen cost per unit of distance, X, by dividing by speed, hence,

$$P_X = \frac{a}{V} + bV.$$

Clearly a minimum is obtained when

$$dP_X/dV = 0, \text{ i.e. when } V_{opt} = \sqrt{\frac{a}{b}}\,.$$

These curves P_V and P_X versus V are shown in figure 1a. However, this optimization is only the first in a series of hierarchical optimization levels, each of which may have subtly different criteria. Thus while energy minimization, or more exactly the minimization of oxygen and/or ATP consumption on an incremental basis seems an appropriate cost function when all other factors are equal, this latter condition is not always so, since there may frequently be a survival value to higher speed, for example, in the predator-prey situation. Here we may model the new total cost function P_T for present purposes by simple functions,

$$P_T = P_V - P_S,$$

where

$$P_S = \text{value function of speed}$$
$$= cV - d$$
$$P_T = (a+d) - cV + bV^2.$$

In this case, the optimal speed is

$$V_{opt} = \frac{c}{2b},$$

and the relation of these cost functions with velocity is shown in figure 1b. Note that the 'constant' charges for zero velocity are of no significance, and that the minimization of cost involves only equalizing the slope of increased energetic cost with the slope of the increased value of speed. Clearly the nature of the analytic functions may generally be more complex than this, and the present ones have only been chosen for simplicity in this review. The main point to note mathematically is that at the optimal speed the incremental cost of speed increase is zero.

Many neuromuscular systems, including those for locomotion, are capable of multi-modal operation and each of these modes may have its own optimal condition of velocity, in the case of locomotion. A more difficult conceptual consideration now involves how these different optima are weighted into one commensurable cost function for the animal as far as an overall evolutionary strategy is concerned. Thus as we have just shown, the optimal 'cruise' extremum may be relevant in the many cases where the animal is in no hurry to go from place to place, whereas at other times the 'maximum power output' extremum may be most appropriate in order to secure maximum velocity. In technology the engineer sadly accepts that it is impossible to design a single system which can achieve multiple optima at the same levels as could be obtained individually; there seems no reason to believe that this situation is different in biology. Thus at a higher hierarchical level another trade-off is involved, in the sense of determining how these different optima should be weighted in evolving the overall 'best' system. Such trade-off obviously must introduce probability considerations as to the frequencies with which the various modes will be needed during the organism's lifetime, and of course, it is exactly such weighting that evolution is able to develop through its large number of parallel experiments carried out over many generations of individuals. This statement of need, however, still cannot inform us as to how the neuromuscular systems are to be programmed neurophysiologically in a co-ordinated way so as to be switched between these different optimal modes when necessary.

Another aspect of basic concern to this muscular co-ordination theme is the evolutionary trade-off between 'brain' and 'brawn'; that is, between increasing informational capacity and increasing muscle power. For example, note that an 'informationally-rich' but muscularly-weak predator, or prey, may achieve his objective at least as well as competing colleagues who might conversely be informationally-poor but muscularly-powerful. To do this he must adopt a better strategy in chase or escape, based upon information processing and prediction so that his effective capability is as high as that of his competitor who is capable of higher speed. Figure 2 illustrates this diagrammatically for a simple chase sequence, but it should be noted that this becomes more complicated mathematically if random type evasive motion is undertaken by the prey. Clearly man is tempted to think that he at least has proven that brain is more powerful potentially than brawn, and indeed, we must agree for our particular predator-prey society. On the other hand, evolution also operates at the ecological level, and here the relevant criteria for ecological optimization primarily seem to relate to maximizing the stability of populations, their probability of survival, and/or maximization of negentropy production. These criteria could operate so as to competitively exclude those ecologies in which particular species sufficiently disturb the stability of the ecology and its population. In particular, man, through his high muscular co-ordination and through his augmentation of its effectiveness by use of tools and weapons, could unhappily provide an example of this in his own ecology.

III. Neuromuscular Control Structure and Strategies

The major structural features of mammalian skeletal neuromuscular systems are probably agreed on, although many refinements have been published in recent years regarding especially the innervation of intra- and extra-fusal muscle fibers, while the subjects of the roles of Renshaw-cell collateral loops, spinal-level interneuronal connections, Golgi tendon organ roles, and 'long-loop' CNS pathways have yet to be definitively modelled. During this recent period it has generally been considered that the neuromuscular system must operate as a classical servo-mechanism, probably with γ-pathway and spindle cells respectively providing the basic reference command and error actuation, as exemplified in the classical stretch reflex.

It is important to note that knowledge of this structure, however, does not enable us to predict the detailed functional operation and strategies of the system, especially because the relative activities of the various higher loops upon the basic spinal cord loops are not well understood. As one example, the eye movement co-ordination system was modelled by YOUNG in 1962 for the first time as a sample-data model, both for its saccadic and smooth pursuit modes. While it is now generally agreed that a sample-data mode is not appropriate for the smooth pursuit system [YOUNG, 1968], this example, nevertheless, demonstrates that sophisticated modes of control can be implemented within the apparently conventional servo-mechanism-like structure.

1. A basic neuromuscular control actuating mechanism. Recently AIZERMAN and ANDREJEVA [1968] have proposed a new mechanism of basic muscular activity to explain their various experimental results. In brief summary, their explanation is that higher CNS control commands the neuromuscular system to produce 'splashes' of activity at aperiodic intervals in a frequency range centered near 10 Hz. These splashes are initiated by a 'Simplest Search Mechanism' (SSM) after a period of muscle relaxation has allowed an extremum-type 'discomfort function' to pass through its minimum and increase by a threshold amount. The neural command for this splash is delivered to many muscles within the body, but the strength of the splash in any muscle is modulated by 'Random Interneuronal Pools' (RIP) near the α-motoneurons in the spinal cord. In turn, the RIP are fed with proprioceptive information by their respective muscles to provide a basis for this strength modulation. As a result of each splash, the muscles are contracted briefly, akin to the basic muscle fiber twitch, but now resulting from an α-MN volley, and the discomfort function is reduced for the situation of concern; for example, reduction of postural 'error' in the task of merely maintaining an upright balance, or of producing certain time functions as in speech and in hitting or pitching a ball in games such as golf, football, baseball, and cricket. When the discomfort rises again sufficiently after completion of the splash, then a new splash is commanded.

Adaptive or plastic behaviour is implicitly achievable by this control mechanism, either by including in the discomfort function the effects of more muscles than normally necessary, or by changing the discomfort function itself. In the first case substitution patterns become possible, such as when particular joints or muscles which are used in the optimal pattern by a healthy person become inactivated or when there are external physical constraints on the movements permitted, for example in manipulating objects in confined spaces. This ability was dramatically shown in AIZERMAN'S experiments with rats, by feeding back as a pain stimulus, the rectified filtered EMG of a particular muscle in a non-linear functional form having a minimal value at some specific EMG strength. The rats were able to sustain their EMG's at the level which minimized the pain stimulus, through utilizing the SSM and splash mechanism, and furthermore were able to track rapidly any movements of the minimum point which might artificially be introduced.

2. *Learning without a teacher*. As already noted, the 'under-specified' configuration for skeletal posture and movement permits many possible muscle co-ordination patterns, and an optimizing concept is necessary to find the best solution. Further we may ask how the neuromuscular system can arrange to 'learn without a teacher' some at least of the various spinal reflex and higher connections necessary for the observed basic structure. Here we imply that it is not practicable for the genetic blueprint to carry all such minute specifications as those regarding dendritic connections, and information-transmission weighting of each synapse, etc. This view seems generally accepted, especially in view of the system's observed adaptive ability to adjust to random neural 'breakage', etc. during the organism's lifetime. Now the human apparently acquires the basic stretch reflexes early in the gestation stage, at which time the only 'teacher' available is that of the maternally-forced disturbance movements on the various limbs *in utero*. GRIFFITH [1963], therefore, postulated that each spinal reflex could be learned by minimizing some performance criterion, or cost function, appropriately related to the local α-MN activity; specifically that each α-MN would minimize the sum of the squares of all incoming presynaptic action potential (AP) firing frequencies into the particular neuron, f_i, plus the square of the output frequency, f_0.

Thus,

$$P = f_0^2 + \sum_{i=1}^{N} f_i^2.$$

One basis for this choice of criterion is probably its mathematical tractability, but the function does seem to represent a plausible trade-off between the energy needed for muscular effort (as represented by the f_0^2, since f_0 produces one-for-one muscular motor unit contraction), and the 'discomfort function' due to error in maintaining the desired situation (as represented, for example, by the Σ-term which may be considered to sum the proprioceptive error signals after being modulated by their γ-efferent set point commands).

This system can be made to 'steep descend' towards a minimum by well-known analytical techniques and for the present we merely note that one conventional basis for this is to adjust the synaptic weights w_i of the ith synapse on the given neuron, by the rule

$$\frac{dw_i}{dt} = -k \frac{\delta f_i}{\delta f_o}.$$

By using approximations to this idealized scheme, GRIFFITH produced learning of a spinal reflex after a few imposed cycles of disturbance torque.

This interesting theoretical work leaves unexplained how the system would actually implement the information flow feedback paths for this optimal-seeking behaviour. However, it is interesting to note that experimental evidence now exists for optimal behaviour in a number of major neuromuscular systems, which in turn makes this approach at the simple reflex arc level appear fruitful. Thus some diverse examples of optimization in major systems are in respiration [OTIS *et al.*, 1950], walking [COTES and MEADE, 1960], swimming of salmon [BRETT, 1965], and flying of birds [TUCKER, 1968]. Finally we note an analytical attempt by NUBAR *et al.* [1961] to relate such optimization ideas to the control of posture, by using the total mean square muscle torque as the performance criterion to be minimized in maintaining any particular posture. At this early stage, such analyses are of course based upon the assumption of equal muscle efficiency in developing given torques.

3. Preprogramming. A very rich feedback structure is available to control neuromuscular activity, from the basic spinal-cord loop up through the long loops in the CNS. One potential advantage of this extra complexity could be realized by incorporating elementary learning and prediction of the neuromuscular activity necessary to achieve the many particular 'skilled' tasks we use in everyday life for locomotion, feeding, manipulation, and especially for producing speech. In particular, since continuous feedback control is only more efficient than open loop control when significant random changes occur continually in the neuromuscular parameters and in the disturbance input, the system could achieve higher dynamic capability for a given capacity of its neuromuscular actuators if specific movements were controlled on an open-loop basis, with the feedback loops operating in the monitoring role, whereby they did not interfere unless unexpected, deteriorated muscular performance and/or its co-ordination occurred; for example, much as the aircraft pilot and machinist monitor their semi-automatic machines. From common observation, there seems little doubt that such preprogrammed control is used in neuromuscular tasks involving considerable muscular co-ordination and frequent repetition. As an everyday example, consider the fast climbing of a set of stairs which all have the same height, except for perhaps an intermediate level platform on which one may well stumble (as on aircraft entry staircases) or except for the last several steps at the top of the staircase (as reputedly in medieval European castles!). Handwriting provides another example in which the visual feedback is only necessary for monitoring the level of the line, etc. Experimentally WATT and JONES [1966] have measured the EMG activity patterns during hopping and jumping manoeuvres of both the voluntary and involuntary kind and shown that the muscles are indeed nicely preprogrammed beyond the capability of at least the stretch reflex, since the muscles are activated as necessary for the landing before it actually occurs.

IV. Muscular Co-ordination Studies at the McGill BioMedical Engineering Unit

The McGill BioMedical Engineering Unit has an experimental program on neuromuscular control mechanisms both at the single joint level, and at the complexly co-ordinated level of posture and locomotion. For the former study, a flexible electrohydraulic actuator has been constructed of adequate bandwidth-torque ability so as to interact with the forearm at its maximum dynamic level. Studies of the control characteristics in both torque and position control mode are in progress, with and without voluntary control, with and without vision, etc. By simultaneous measurement of relevant EMG and dynamic variables, it is hoped to differentiate the influences of some of the various long loop pathways and so to construct meaningful models of this neuromuscular system's control operation.

For the posture study a horizontal platform has been constructed, approximately one meter square, and instrumented with six load cells which measure forces and torques in the six linear and rotational degrees of freedom. On-line instrumentation is used to process these data into six directly useful variables, including total vertical force and center of pressure. Both normal subjects and patients with neuromuscular disorders are being stud-

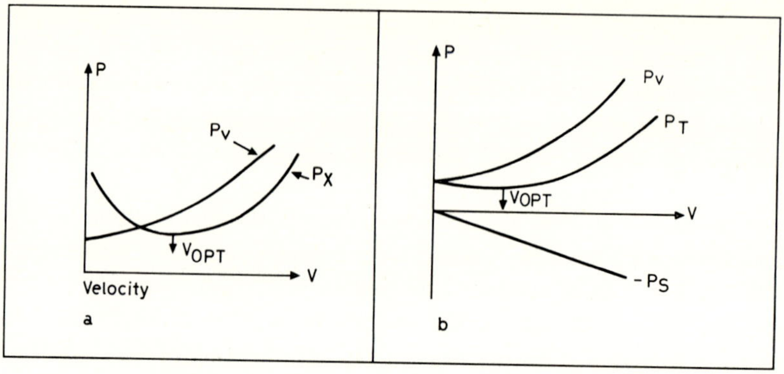

Fig. 1. Optimal cost functions.

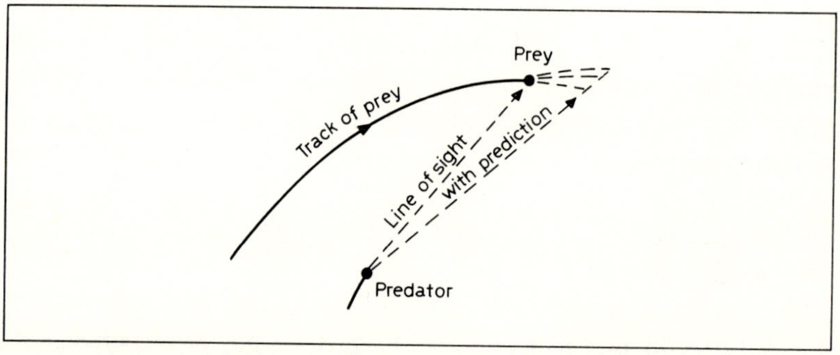

Fig. 2. The brain versus brawn problem.

ied. In the former case it is intended progressively to eliminate degrees of freedom, such as flexure at the knees, hips and neck, so that adaptation to different co-ordination programs will be required between successively more restricted muscle groups in maintaining posture. This also requires that we study the nature of the pertinent performance criteria being optimized under these different conditions, especially the possible role of inertial stabilization of the head by vestibular signals, in view of the head's importance in the visual stabilization mechanism. The instrumentation involves measuring pertinent EMG, and the postural kinematics by co-ordinated cine-photography.

The locomotion study has so far been mainly theoretical and concerned with the necessary torque program to reproduce observed kinematic patterns during the walking gait, but some preliminary treadmill studies have been made.

References

AIZERMAN, M. A. and ANDREJEVA, E. A.: Some control mechanisms of skeletal muscles (in Russian and English). Inst. Automat. Telemech., Moscow (1968).

BRETT, J. R.: The swimming energetics of salmon. Sci. Amer. *213* (2): 76–84 (1965).

COTES, J. E. and MEADE, F.: Energy expenditure and energy demand in walking. Ergonomics *3* (1960).

GRIFFITH, V. V.: A model of the plastic neuron. IEEE Trans. med. Electron., Mil-7 *293:* 243 (1963).

MILSUM, J. H.: Biological control systems analysis (McGraw-Hill, New York 1966).

NUBAR, Y. and CONTINI, R.: A minimum principle in biomechanics. Bull. mathem. Biophys. *23* (1961).

OTIS, A. B.; FENN, W. O., and RAHN, H.: Mechanics of breathing in man. J. appl. Physiol. *2:* 592 (1950).

TUCKER, V. A.: Respiratory exchange and evaporative water loss in the flying budgerigar. J. exp. Biol. *48:* 67–87 (1968).

WATT, D. and MELVILL JONES, G.: Observations on the neuro-muscular control of purposeful movements. Proc. Ann. Sci. Meet. Aerospace med. Ass., Bal Harbour (1968).

YOUNG, L. R.; FORSTER, J. D., and HOUTTE, N. VAN: A revised stochastic sampled-data model for eye tracking movements. 4th NASA-Univ. Conf. Manual Control. Univ. of Michigan, Ann. Arbor (1968).

Author's address: Dr. JOHN H. MILSUM, BioMedical Engineering Unit, McGill University, *Montreal, Quebec* (Canada)

Medicine and Sport, vol. 6: Biomechanics II, pp. 72–76
(Karger, Basel 1971)

The Neuromuscular Control of the Mandibular Movement

F. BOSMAN and M. F. C. VAN GOOL

Department of Medical and Physiological Physics and Dental Institute,
Rijksuniversiteit Utrecht, Utrecht

It is well known that the mandibular movement is important in several oral functions: e.g. speaking, chewing, swallowing. In comparison with other movements it shows some unique aspects. This movement is made possible by a bipartite joint, which requires a complex coordination of the right and the left muscles. Besides, the elevators and depressors of the mandible are rather unequal. Concerning this last point it must be stated that there is not only a difference in possible force between these muscle groups, but also a difference in number of muscle spindles [MURPHY, 1967]. In our approach we try to get information about the neuromuscular control by studying the performance of the system in several conditions.

In this kind of research often a useful distinction is made between preprogrammed movements and movements with feedback [DENIER VAN DER GON and WIENEKE, 1969]. In the latter a continuous correcting control is supposed to be exerted while the control and correction of preprogrammed movements takes place after them. The idea of mandibular movements with some reflex control is generally accepted [KAWAMURA, 1964, 1967, 1968; JERGE, 1964; GRIFFIN, 1965; MURPHY, 1967; HONÉE, 1968]. However, in this case visual feedback is unusual. Nevertheless, it may be of interest to perform mandibular movements with visual feedback and to compare the results with e.g. hand movements. The mandibular movement is then a part of a servo loop with visual feedback.

In our experiments a subject was sitting upright in a chair. On the screen of an oscilloscope he observed a horizontal line moving only in vertical direction. With an electronic device the position of a certain point of the mandible with respect to the maxilla was recorded and made visible as a spot on the same oscilloscope screen. The task of the subject was to track the moving line by opening and closing his mouth. As a variation on this tracking task we eliminated the visual feedback by extinguishing the intensity of the position spot. Then no visual perception of the difference between target and position was possible. In order to avoid movements as a result of pre-

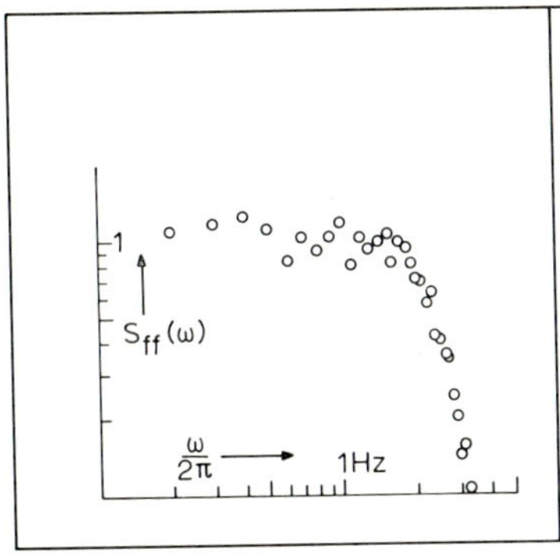

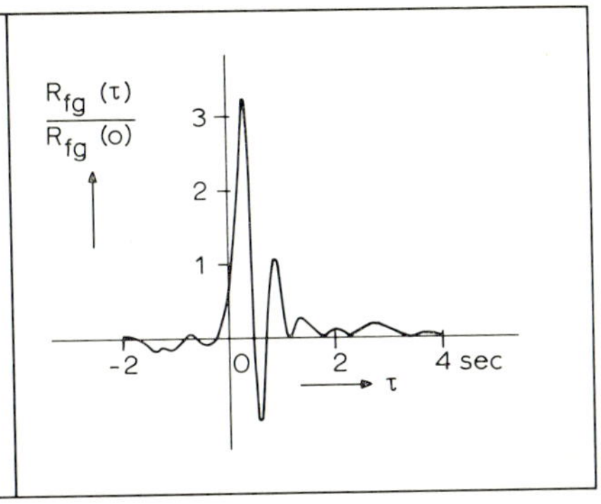

1, 2

Fig. 1. Power spectrum of a target signal.

Fig. 2. Normalized cross correlation function of the target signal and the tracking signal.

diction the target moved unpredictably: the input signal was white noise within an adjustable bandwidth (fig. 1). Besides the position, the velocity and acceleration could also be recorded.

The performance of this servo system was described with the help of correlation functions and power spectra. Figure 2 presents the cross correlation function $R_{fg}(\tau)$ of the target signal f(t) and the tracking signal g(t) (i.e., the position of the jaw). $R_{fg}(\tau)$ achieves a maximum $\hat{R}_{fg}(\tau)$ for τ 200 to 250 msec. This suggests a delay between these two signals. The results were in a good agreement with those found in reaction time measurements.

A comparison was made between tracking by moving the mandible and tracking by moving the hand rotating in the wrist joint. As a measure of the quality of tracking we took the normalized maximum value of the cross correlation function according to:

$$K = \frac{\hat{R}_{fg}(\tau)}{\sqrt{R_{ff}(0)\, R_{gg}(0)}}.$$

R_{ff} (0) and R_{gg} (0) are the maxima of the autocorrelation functions of the target and the tracking signal respectively. K lies between 0 (= no correlation at all between the two signals) and 1 (= the tracking signal is identical to the target signal, apart from the delay). It is not surprising that K depends on the bandwidth of the target signal. Figure 3 represents K as a function of this bandwidth for hand and mandibular movements. It is remarkable to note that K is about the same for hand and mandible. This suggests that

3, 4

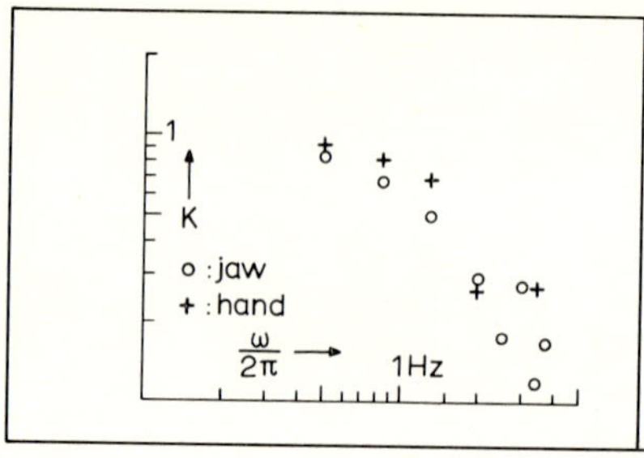

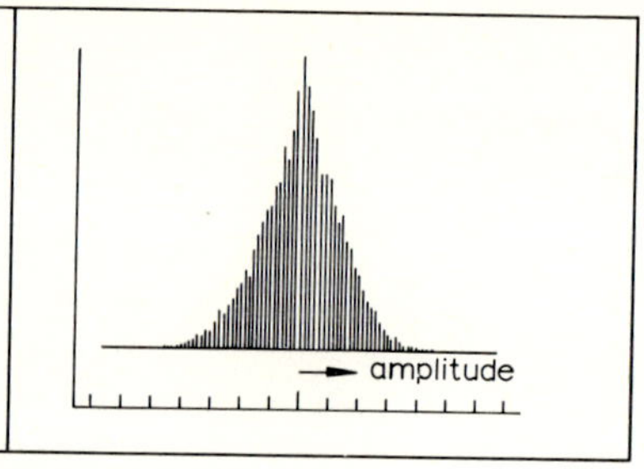

Fig. 3. Tracking quality K as a function of the bandwidth of the target signal.
Fig. 4. Amplitude distribution of the velocity of a tracking signal.

the limits of both servo systems are not situated in the muscle limb system, but have a more central common origin.

Several studies in this field suggest the existence of a '*sampled data system*' for eye and hand movements. According to this theory the target signal is perceived in samples and the resulting tracking signal is stepwise. The arguments for this hypothesis are based on several experiments. With this hypothesis in mind we calculated the amplitude distribution of the tracking signal. In our experiments the input signal was white noise, limited in the frequency range. It is well known that the amplitude distribution of such a signal as well as that of its first derivative is gaussian. When the subject tracks stepwise, the amplitude distribution of the tracking signal may be of the same character as that of the target signal, but within the velocity distribution of the tracking signal some preference for higher velocities can be expected. Figure 4 shows a velocity distribution. The bandwidth of the target signal is 1 Hz. This figure does not show a striking deviation from a gaussian distribution. The same negative result was found for other target signals with various bandwidths. Concerning the last check, however, it is possible that the inertia of the jaw gives a smoothing effect, resulting in absence of fast changes in position. In any case, it did not confirm the data sampling hypothesis.

An argument in favour of the sampling mechanism is the presence of a peak in the frequency response of the tracking system. Navas and Stark

5, 6

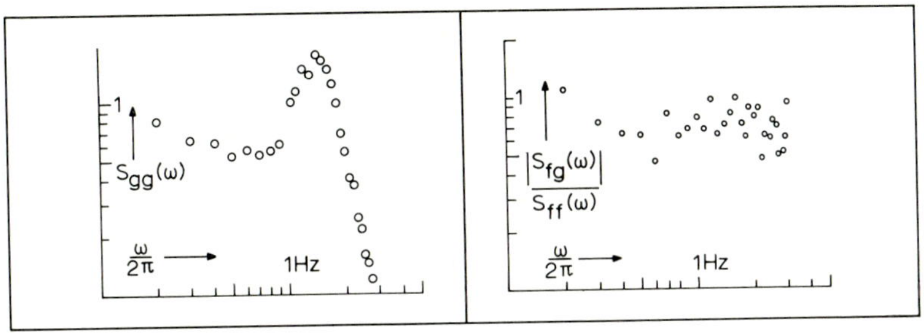

Fig. 5. Power spectrum of a tracking signal.
Fig. 6. Magnitude of a transfer function.

[1968], dealing with hand control systems, call for instance attention to this peak appearing at 1.5 Hz. In our experiments with visual feedback the power spectra of the tracking signals showed a peak too (fig. 5). However, the position of this peak was changing with the cut-off frequency of the target signal in nearly all cases, and could be as high as 3.5 Hz. If this peak is the result of some kind of intermittency, it means that the sample frequency depends on the target signal. If the bandwidth of the target signal is larger than 1.5 Hz, the quality K decreases rapidly. If the peak is the result of a sampled data mechanism and the peak frequency a measure for the sampling rate, the increased sampling rate for target signals of higher bandwidth does not contribute. So the energy, condensed in the peak, has no tracking quality. This is confirmed by calculating the transfer function $\Phi(j\omega)$ from

$$\Phi(j\omega) = \frac{S_{fg}(j\omega)}{S_{ff}(j\omega)},$$

where $S_{fg}(j\omega)$ is the cross power spectrum. According to this procedure figure 6 gives the magnitude of $\Phi(j\omega)$. It shows hardly any preference for some frequency.

Eliminating visual feedback results in a much more pronounced peak in the output power spectra. The origin of the peaks is still under discussion and a subject for experiments. The fact that the peaks have no tracking quality may mean that the subjects make nearly periodical movements, triggered by the fast changes of position of the target. This effect is accentuated when there is no visual control. It may be suggested that the tracking system consists of more than one superimposed mechanism for data processing, one of these being the system without visual feedback.

In the preceding part attention was payed to the differences between the elevators and the depressors. The depressors are seen to possess fewer spindles than the elevators. So the present system is not symmetric and it should be considered as a *'rate sensitive system'*. However, in the experiments mentioned above, no significant difference between opening and closing movement has been observed. It may be expected that differences appear when forces act upon the mandible. We are developing an apparatus with the help of which it will be possible to make pulse-like and continuous forces act upon the mandible. In this way reflex activity of elevators and depressors can be investigated as well as the mechanical output impedance of the system.

References

DENIER VAN DER GON, J. J. and WIENEKE, G. H.: The concept of feedback in motorics against that of preprogramming. Biocybernetics of the control nervous system, pp. 287–296 (Little, Brown, Boston 1969).

GRIFFIN, C. J.: Inhibition of the linguo-mandibular reflex. Austr. dent. J. *10:* 376–383 (1965).

HONÉE, G. L. J. M.: Enige reflexsystemen in het kauwstelsel. Ned. T. Tandheelk. *6:* 451–458 (1968).

JERGE, C. R.: The neurologic mechanism underlying cyclic jaw movements. J. prosth. Dent. *14:* 667–681 (1964).

KAWAMURA, Y.: Recent concepts of the physiology of mastication. Advances in oral biology, vol. 1, pp. 77–109 (Academic Press, New York and London 1964).

KAWAMURA, Y.: Neurophysiologic background of occlusion. Periodontics *5:* 175–183 (1967).

KAWAMURA, Y.: Mandibular movement: normal anatomy and physiology and clinical dysfunction. Facial pain and mandibular dysfunction, pp. 37–82 (Saunders, Philadelphia 1968).

MURPHY, T. R.: Shortening/inhibition of prime movers. Brit. dent. J. *123:* 578–584 (1967).

NAVAS, F. and STARK, L.: Sampling or intermittency in hand control system dynamics. Biophys. J. *8:* 252–302 (1968).

Authors' addresses: Dr. F. BOSMAN, Dept. of Gnathology, Dental Institute, c/o Dept. of Medical and Physiological Physics, Rijksuniversiteit Utrecht, Eisenhowerlaan 4, and Dr. M. F. C. VAN GOOL, Dept. of Medical and Physiological Physics, Rijksuniversiteit Utrecht, Eisenhowerlaan 4, *Utrecht* (Netherlands)

Medicine and Sport, vol. 6: Biomechanics II, pp. 77–82
(Karger, Basel 1971)

Analysis and Synthesis of Handwriting

W. G. Koster and J. Vredenbregt

Institute for Perception Research, Eindhoven

1. Introduction

An everyday example of muscle co-ordination and of a controlled motor activity is handwriting. It is a subject on which remarkably little research has been carried out. Attempts to use an automaton to recognise cursively written texts have hitherto been unsuccessful. This is less likely to be due to technical incompetence than to an insufficient knowledge of human production and perception of this mode of language expression. Only a limited number of studies have been made on the production of handwriting [Denier van der Gon and Thuring, 1965; McDonald, 1965; Eden, cited by Lindgren, 1965]. The above considerations, together with the knowledge gathered from studies on muscle mechanics [Vredenbregt and Koster, 1966], prompted us to undertake the study reported here.

2. Analysis

To gain more insight into the way people write, we started a study in which pencil movements were recorded during the writing of separate letters.

When a letter is written, for instance an 'a', pencil movements are controlled by several muscles. These movements may be broken down into two mutually perpendicular directions.

The x-movement, in the direction of writing, is mainly the result of a rotation of the hand from the wrist or a rotation of the forearm from the elbow joint, the y-movement is largely produced by the thumb and the first two fingers. Actually, movements are also made in a third direction to put the pencil on to the paper and to lift it.

To analyse the pencil movements during cursive writing an instrument was constructed which could measure the amplitude of the movements in the x- and y-directions as a function of time. The displacement-time diagrams were recorded. For separate letters the movements in the y-direction are larger than those in the x-direction. In a preliminary study it was found that the two movements gave comparable results. In the experiments to be described here, only movements in the y-direction have been investigated. In this article we will use the letter 'a' by way of example. The studies have been carried out on various letters of the alphabet.

While different subjects wrote their own special letter 'a', the displacement-time diagrams were recorded. Individual differences in the shape of the diagrams were apparent. The diagrams were then analysed. Characteristic points in the diagrams are the moments at which the direction of movement is reversed. The time intervals between successive moments will be called t_1, t_2, t_3, t_4 and t_5 (fig. 1).

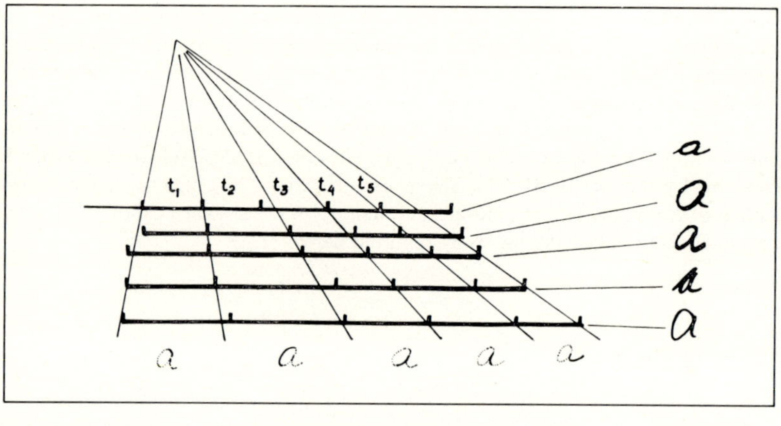

Fig. 1. The letter 'a', as written by five subjects, together with the successive interval times t_1–t_5. In this figure a constant ratio between the interval times, independent of the subjects' individual way of writing, is indicated.

The total writing time differs considerably from one subject to another. The ratio of the time intervals is seen to be fairly independent of the subject's individual way of writing. The same result is obtained when one subject writes several letters 'a' either quickly or slowly or when he writes a large or small 'a'. The number of time intervals and the ratio between them seem to be characteristic of the letter of the alphabet.

Up to this point we have spoken only about the movements that produce a letter. These movements are controlled by various groups of muscles and it would be of interest to have detailed knowledge of the activity of each of these muscles during the writing of a letter. During muscle activity electric signals generated in the muscle may be displayed in the form of an electromyogram. A preliminary inspection of the recording of the electromyograms revealed varying periods of muscle activity. Although the technique used for picking up electromyographic activity by surface electrodes is not ideal in this case, we were able to make the following observations. The curves representing the electromyographic activity in relation to the displacement-time curve during writing show that, just before the direction of the movement is changed, the muscles concerned become active. The activity disappears when the pencil-point has attained the appropriate speed. After that, the pencil moves further by the inertia of the hand until the activity of the antagonist muscle changes the direction of the movement. It could not be established whether the levels of activity were constant. For detecting such levels, as well as for accurate determination of the points of change and durations of the electromyographic activity, other techniques will have to be used.

A written letter may be thought of as being generated by a complex pattern of short periods of constant activity in agonist and antagonist muscle groups, working in two perpendicular directions and each causing a force to be built up. In order to test whether such a hypothesis is tenable and to be able to introduce voluntary changes in the time patterns, a simulator was developed – a mechanical imitation of the hand. It comprises several characteristics of the hand-arm system such as damping and elasticity.

3. Synthesis

Skeletal movement is brought about by two groups of muscles. Activity in the one, the agonist, causes a limb to rotate around its joint in one direction; activity in the other, the antagonist, causes a movement in the reverse direction. When the agonist contracts, the antagonist is passively stretched. The same principle is used in the simulator (fig. 2).

The simulator consists of two pairs of d.c. electric motors, one pair mutually coupled for movement in the x-direction, the other pair, which is also mutually coupled, for movement in the y-direction.

Fig. 2. The handwriting simulator.

When a particular motor is actuated, its cross-coupled partner operates as a generator of electricity. This energy is dissipated in a resistor providing damping of the system.

In the analysis, short periods of activity were measured. The same principle is copied in the simulator, the electric motors being actuated by pulses of certain duration. The timing of the pulses is controlled by a special-purpose computer. The voltage of the pulses is kept constant.

A normal pencil is used, and just like the hand the simulator may lift the pencil off the paper and move it some distance in order to be able to write the next letter. In this way it proved to be possible to write letters that are like normally written ones. The total writing time and the displacement-time diagrams are very similar to those of the handwritten letter.

As in muscles, excitation is switched on just before the direction of movement is changed. When the terminal voltage is switched off, the inertia of the system causes the movement of the pencil-point to continue.

Slight changes in the timing programme only lead to different shapes of the letter 'a', each being characteristic of one individual (fig. 3). As we are not yet able to synthesise a letter on the basis of rules, the timing pro-

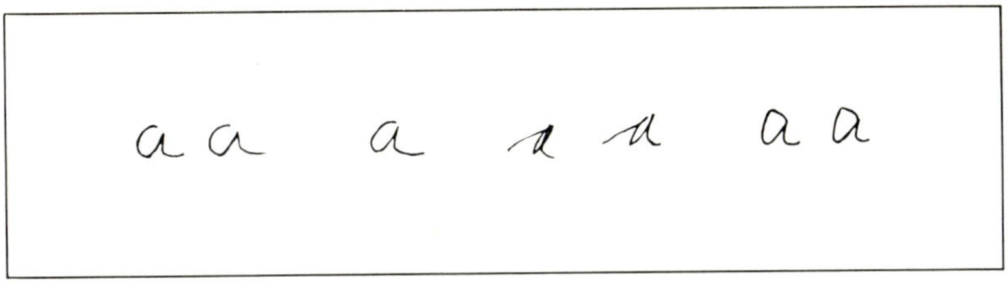

Fig. 3. Imitations of letters 'a', as shown in figure 1 written by the simulator.

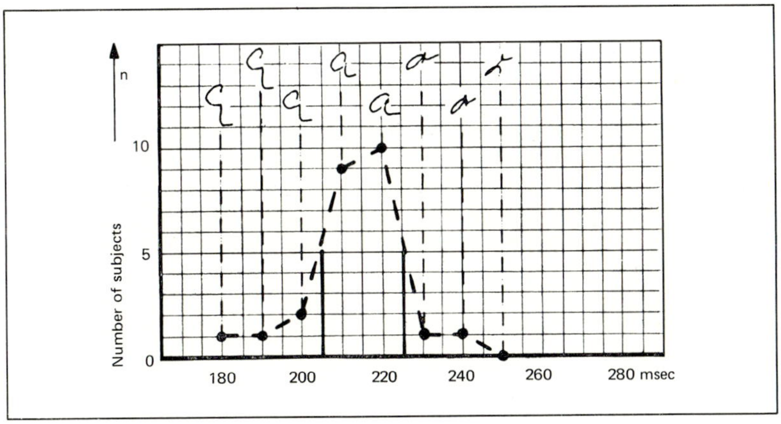

Fig. 4. The figure shows the result of alterations of the moment a particular pulse is initiated. The number of subjects that accepted a particular character as an 'a' is plotted as a function of the moment the pulse is initiated. The different shapes of the letter 'a' are indicated as well.

gramme is made by trial and error. The result is judged by means of perceptual analysis.

The influence of changes in the programme on the perceived shape of the letter has been studied systematically. Each pulse is characterised by its position on the time scale and by its duration. These parameters were varied and the corresponding letters were presented to ten subjects. They were asked to indicate which character could be accepted as a particular letter in a written word. With this technique we took into account to some extent the considerable variations in the shape of a letter that are accepted in cursive writing (fig. 4).

The number of subjects that accepted a character as a natural 'a' was plotted as a function of the position of a pulse in the timing programme. The difference between the 50% points are a measure of the tolerances in timing. Thus it was found that the tolerance of a shift in position of a particular pulse of fixed duration along the time axis is twice the tolerance of an alteration of the duration of that pulse. Particular pulses in the timing programme were found to have much smaller tolerances than others.

Generally speaking it may be said that changes in the timing programme of 5 msec are hardly perceptible. Changes of 20 msec will often make a character unrecognisable.

4. Conclusion

The basic hypothesis that the movements during handwriting are caused by four groups of muscles, each being active during relatively short periods of time, is found to be in agreement with the results obtained with the simulator. It proved to be a useful basis to work on. The timing of the pulses had to be very accurate in order to produce acceptable script. An interesting side-result was that individual script could be obtained by slight changes in the timing programme only.

References

DENIER VAN DER GON, J. J. and THURING, J. PH.: The guiding of human writing movements. Kybernetik *2:* 145–148 (1965).

LINDGREN, N.: Machine recognition of human language. III. Cursive script recognition. IEEE Spectrum *2:* 104–116 (1965).

MACDONALD, J. S.: Investigation of handwriting signals. Quart. Progr. Rep. MIT *76:* 210–227 (1965).

VREDENBREGT, J. and KOSTER, W. G.: Some aspects of muscle mechanics *in vivo*. JPO Annual Progress Report *1:* 94–100 (1966).

Authors' address: W. G. KOSTER and J. VREDENBREGT, Institute for Perception Research, Insulindelaan 2, *Eindhoven* (Netherlands)

Medicine and Sport, vol. 6: Biomechanics II, pp. 83–91
(Karger, Basel 1971)

Motor Subsystems

M. Nemessúri

Hungarian Institute of Physical Education, Budapest

At the first International Seminar of Biomechanics our investigations on human muscles acting by forming three operator pairs were reported. We supposed that the main dynamic effect, based on postural reactions, is pulsion and traction, pushing and pulling, which alternate periodically.

The present study concerns the data obtained from further observations regarding pulsional and tractional moving actions. We would suggest that in the innumerable moving patterns realizable by the human motor system [3] only three moving elements are traceable, viz. one obligatorily static and two dynamic elements in an alternating sequence of pulsion and traction. Connected in parallel and in series, these sequences form the simple and the complex moving patterns.

To examine our suggestion we analyzed different moving actions as walking, running, shot-put, throwing and catching a ball, weight lifting, and other motor actions. These analyses were carried out with the aid of electromyography (EMG) and photokinography.

We studied those basic motor mechanisms and their reflections on EMG [1, 5, 6] and photokinography [4, 5] respectively, which were recorded during each movement in order to determine its basic elements.

We started from the idea that 3 types of motor organisations could be distinguished from which the motor program and its realisation, viz. the motor pattern, are recordable and measurable. After obtaining these two factors we were able to draw conclusions regarding the third, which is the motor sketch stored in the central nervous system (CNS).

During the repetition of a movement, the learning process acquired, e.g. by the feedback information of the muscle perception, realizes the moving pattern as well as its moving program and thus corrects the motor process [2].

In the movements analyzed we defined the moving elements or motor elements as the results of static and dynamic muscle actions, in which the location of the acting muscles, their force and their spatial and temporal processes are recordable and measurable, and distinguishable from the preceding and following simple moving processes [6] .

Walking is a very suitable motor process to eliminate the characteristics of separate moving elements. It is one of the most typical human movement patterns and is studied by many investigators. Most scientists consider walking to be a complicated process in spite of its apparent simplicity. According to most investigators the three main periods of walking are: pushing off, swinging, and supporting. We think that the swing cannot be understood as a basic element of motor activity since it is the characteristic moving form of walking. But the concept changes when we convert the swinging phase into two components for functional and morphological reasons: pushing and pulling/pulsion and traction.

Figure 1B shows that in the knee joint of the front leg pushing is a good measurable and limited moving element. Simultaneously, the back leg (fig. 1A) pushes forward the whole body. It is the main power phase of the whole movement which we call the macrodin period.

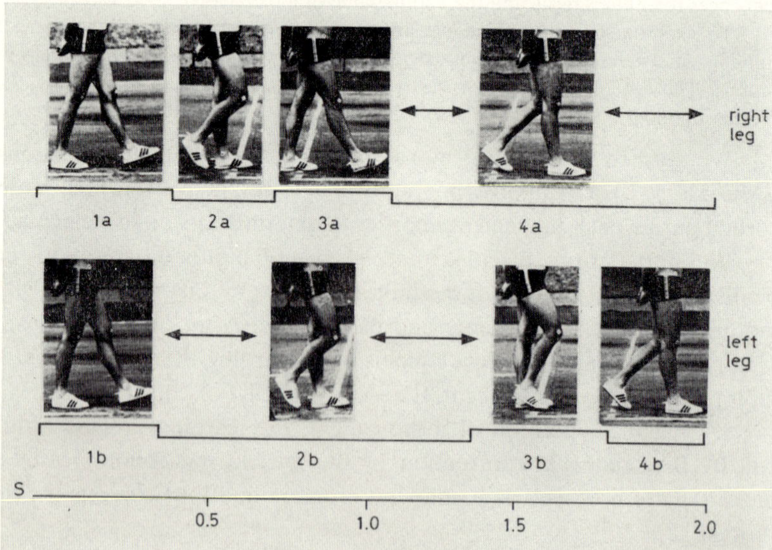

Fig. 1. Motor pattern of walking.

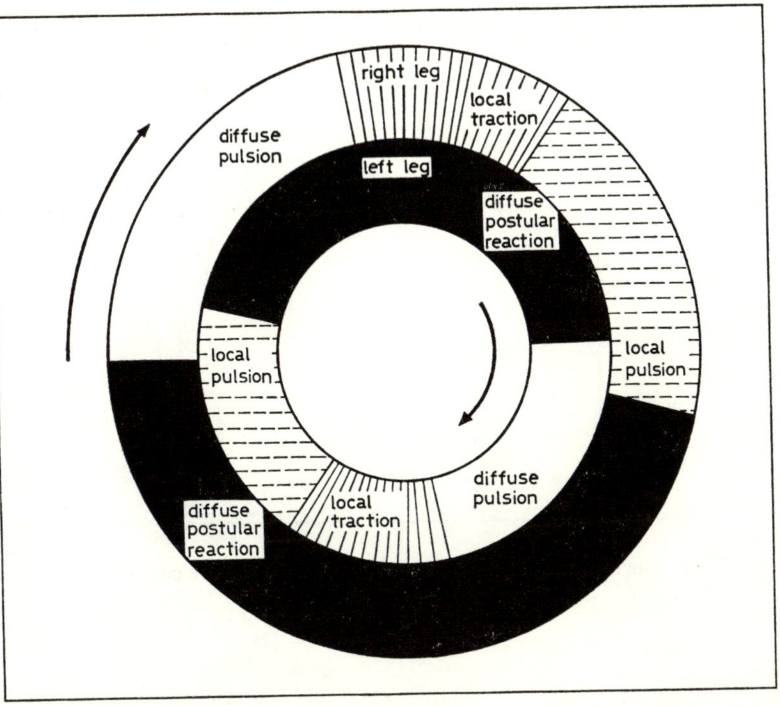

Fig. 2. One cycle of the walking process.

The moving element in figure 2A is pulling the leg which is in backward position (flection of the knee). This phase is followed by lifting the back foot (traction); next, the foot in the backward position is pushed forward to just before the supporting leg (fig. 3A). In figure 4A the moving element of this leg now in the foreward position, is the supporting phase (postural reaction). The sketched moving elements constitute together a part of the whole motor act. The moving periods of the leg in backward position are also represented in figure 1 (1B–4B).

One cycle of the total walking process is represented in figure 2, in accordance with the sequences of the main moving elements of the lower extremities. In the outer circle is represented the power phase of the leg concerned, the macrodin period (in white). The moving element is the diffuse pulsion, i.e. pushing off from the ground. This phase is followed by two further dynamic moving elements. They are the pulling and pushing forward of the leg, followed by the static moving element, the

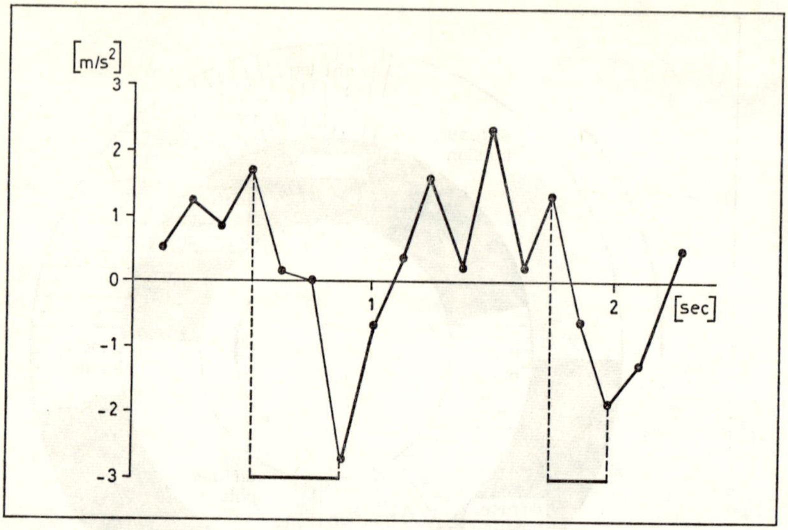

Fig. 3. The influence of the four moving elements on the centre of gravity.

supporting phase after the landing of the foot on the ground (postural reaction). The last phase is the longest period of the total cycle, almost half of it.

The process of movement of the leg in front position is represented in the inner circle, where also four similar moving elements follow each other. The function of the front and the back positions of the leg changes when the leg in backward position takes up the supporting role at the moment when the position of the front leg is changed into the backward position.

The pushing-off phase (the macrodin-pulsion phase) occurs a short time before the other leg's postural reaction starts; thus for a short period the balance of the body becomes unstable during walking. Similarly, the leg in front position moves to backward position, and inversely. Meanwhile, four moving elements follow each other. They are pulsion, traction, pulsion, and postural reaction. These moving elements are connected in series for each limb. They are well distinguishable from each other and are different with respect to space and time according to the second figure, and their dynamic behavior is precisely measurable.

The influence by the four moving elements on the centre of gravity (CG) of the body is a further important datum and is represented in fig-

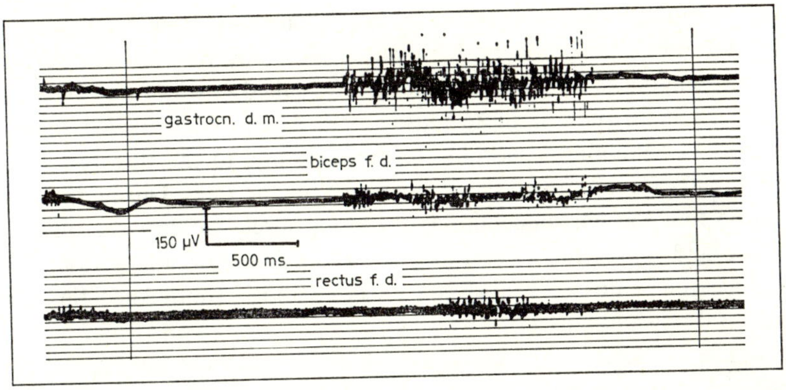

Fig. 4. Pulsors of the hip (m. biceps femoris), the knee joint (m. rectus femoris), and the leg (m. gastrocnemius).

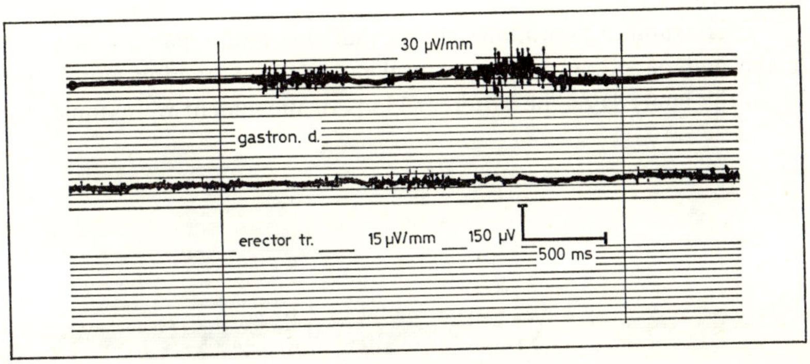

Fig. 5. Pulsor of the trunk (m. erector spinae) acts alternately with the pulsor of the leg.

ure 3. A diffuse pulsion, pushing off, causes the increase of the speed of the CG in the direction of walking, while the local traction and pulsion leads to deceleration.

In figure 4 the muscle power is localized by EMG. During the main power release (the macrodin-motor element) the following muscles are strongly active: gastrocnemius, biceps and rectus femoris, visualized by the sequence of the electrical activity. According to figure 5, the erector spinae and the gastrocnemius act alternately. The glutaeus maximus and the quadriceps also belong to this series of muscles. The muscles mentioned exert the main power during the pushing phase and they play a part during the supporting phase too. They are, therefore, called postural-pulsion muscles.

The main part of these muscles are extensors, in which the biceps femoris – belonging, according to the anatomy, to the flexors – achieves the greater force through the extension of the hip joint. During the bending of the foot the gastrocnemius is the first muscle to start the pushing-off phase and represents a force of diffuse pulsion. For this reason, we call the gastrocnemius as well as the biceps femoris (both knee flexors) the postural-pulsion muscles.

The EMG investigation shows that the adductor longus, the tibialis anterior and other muscles which take part in the flexion, approaching and turning of the hip joint, are also activated, but this activation appears later than that of the postural-pulsion muscles. Their activity is small and lasts a short time (figs. 6 and 7). We think that they belong to the traction muscles.

This example of walking shows that this motor pattern can attribute to the understanding of the postural pulsion and traction-motor systems. From the analysis of the different kinds of movement we assume that the

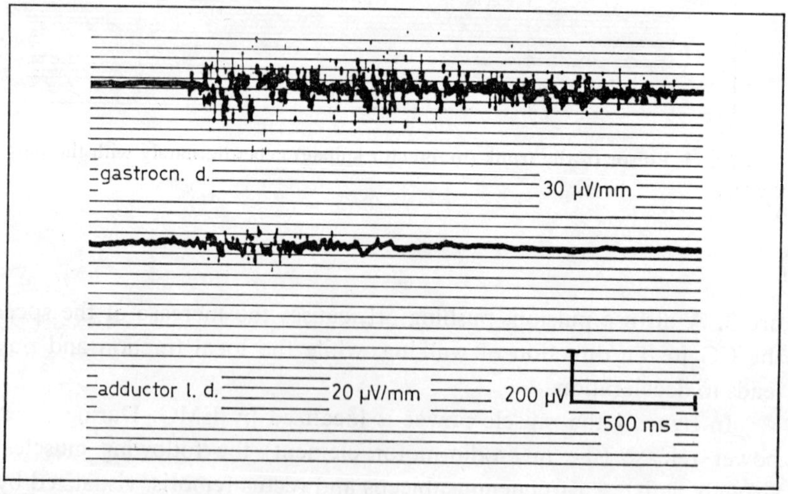

Fig. 6. The pulsor of the leg (m. gastrocnemius) has a great amount of electromyographic activity in walking, the tractor of the hip joint has a small.

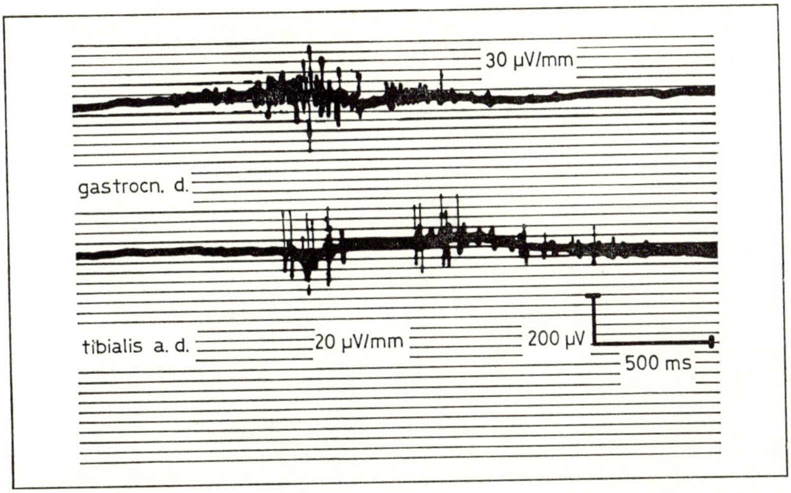

Fig. 7. As in figure 6, the pulsor of the leg has a greater electromyographic activity in walking than the tractor of the same part.

extensors, abductors and exterorotators belong generally to the postural-pulsion muscles (fig. 8). For this reason we judge the function of the muscle according to the main force and not according to the traditional anatomical classification.

On the ground of similar principles we believe that the flexors, adductors and interorotators belong to the traction muscles (fig. 8). On the analysis of the functional organisation of motor elements during walking and other moving processes it might be possible to characterize in detail the main peculiarities of the suggested two motor subsystems.

We have seen that the greater part of the motor patterns during walking is built up from postural-pulsion-motor elements and they form the macrodin period. The analysis of other motor patterns suggests also that the periods of power are mainly constituted by this motor subsystem. The postural-pulsion muscles form the greater part of the muscles of the body, especially in the lower limbs where their ratio is 4.5:1 [7]. Their importance is accentuated by the fact that every movement is based on postural reactions and they are achieved by the postural-pulsion system. The main direction of movement, force, speed and rhythmicity of motor pattern is in the greater part of motor events achieved by the postural-pulsion motor system in the lower limbs.

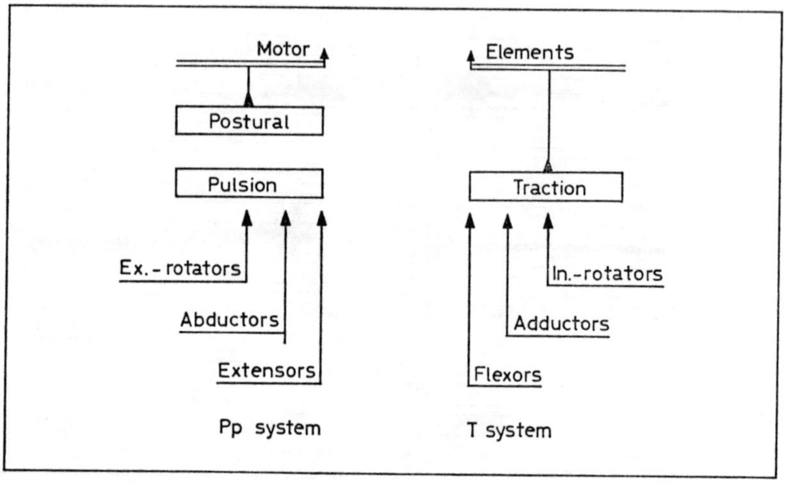

Fig. 8. Binary activity of the motor system.

What could be the role of the other motor subsystem which we have called the traction system? The EMG and other investigations show that they generally belong to the antagonists during the macrodin period. During this period they are stretched and are continuously sending information to the CNS with the aid of the excited muscle spindles. This process plays an important part in the feedback function during learning and improving our movements. On the other hand antagonistic activity accomplishes the function of a second regulating mechanism in the motor function. For this reason our investigations suggest that the motor patterns of directed movements are issued from ready motor programmes stored in the CNS. Motor programmes should be constructed from three motor elements. They are one obligatorily static, postural element and two alternating dynamic motor elements, viz. pulsion and traction. They build the simple and complex movement patterns in a extremely well-combined manner, coordinated in parallel and in series.

However, the postural reaction and the pulsion are achieved by the same motor system, and our conclusion might be that there exist two motor subsystems: a postural-pulsion and a traction system. The postural-pulsion system achieves generally the postural reaction and the macrodin period of movement. The traction system acts as a second regulating mechanism in the total of the muscular work and supplies uninterrupted information by feedback from the stretched muscles.

References

1 BASMAJIAN, J. V. and LATIF, A.: Integrated actions and functions of the chief flexors of the elbow. J. Bone Jt Surg. *39:* 1106–1118 (1957).
2 HEBBELINCK, M.: Ergometry in physical training research. J. Sports Med. phys. Fitn. *9:* 69–70 (1969).
3 JOKL, E.: Physiology of exercise (Thomas, Springfield 1964).
4 NEMESSÚRI, M.: Funktionelle Sportanatomie (Akad. K., Budapest 1963).
5 NEMESSÚRI, M.: L'applicazione dell'analisi mediante la cinematografia ultrarapida del movimento nei varii sports. Ung. oggi *6:* 46–54 (1965).
6 NEMESSÚRI, M.: Der binäre, antagonistische Mechanismus der Bewegungssteuerung. Biomechanics I. 1. Seminar, Zürich 1967, pp. 165–171 (Karger, Basel 1968).
7 SCHUMACHER, G. H. und WOLFF, E.: Trockengewicht und physiologischer Querschnitt der menschlichen Skelettmuskulatur. I–II. Anat. Anz. *118:* 317–330; *119:* 259–269 (1966).

Author's address: Dr. MIHALY NEMESSÚRI, Magyar Testnevelési Föiskola, Alkotas Utca 44, *Budapest 114* (Hungary)

III. Motor Learning Process

Medicine and Sport, vol. 6: Biomechanics II, pp. 94–99
(Karger, Basel 1971)

Cognition and Motor Learning

B. J. CRATTY

Perceptual-Motor Learning Laboratory, University of California, Los Angeles, Calif.

A number of theoretical positions relating thought and movement may be found in the philosophical, historical and experimental literature. The philosophers of ancient Greece evidenced a mind-body monism which was reflected in their statues of the human figure on which shrewd faces were placed above well-muscled torsos.

The Christian Idealist of the pre-Renaissance period espoused a separatism of mind, body and spirit. Educators today who embrace this same philosophy are usually found to be inserting physical activity into their programs only to be dispelling the muscular energies that prove 'harmful' to the most important job of the school, the development of the mind.

In the 18th and 19th centuries the pendulum swung in the other direction. PESTALOZZI, ROUSSEAU and others developed educational programs which were based upon the supposition that a child's mind and body interacted as a unity. The organismic and biosocial theories outlined in more recent times by OLSON [1959] and GARDNER MURPHY [1958] are contemporary statements of this monistic approach.

Movement and cognition may be discussed in more pragmatic terms. Some have attempted to discover just how the intellectual processes contribute in the development of physical skills while, conversely, others have attempted to ascertain just how motor activities may contribute to the development of the intellect [CORDER, 1965; CRATTY, 1969; MACCOBY, 1965].

Experimenters and theoreticians dealing with these two diverse but related questions have sometimes been guilty of oversimplification and/or misinterpretation of their findings. For example, in the American literature on the subject, some have suggested that movement is the basis of the intellect when low non-predictive correlations are obtained between complex movement tasks (requiring the chaining together of a number of responses) and scores obtained from intelligence tests (containing tasks whose successful execution also depends upon serial memory ability [ISMAIL *et al.*, 1967, 1968]).

Others have claimed to have raised the IQ of retardates by exposing them to movement experiences, when in truth, the personality of the experimenter and the extra attention he afforded the subjects were probably more influential of change than was the program of motor activities.

I will advance two hypotheses for your consideration today: (1) Movement will aid a child to think to the extent to which thought accompanies his movements; (2) Thought will aid skill to the extent to which cognition is encouraged at opportune times within the learning process.

A study of bilateral transfer contains evidence that socalled 'motor learning' might be termed 'cognitive motor learning'. A group of human subjects received practice in a onehanded manipulative task, and then were asked to perform the same task with the other hand. As expected, bilateral transfer occurred, skill acquisition by the second hand was apparently facilitated by the activities of the initial practice. The data revealed that as much skill was evidenced by the second group of subjects who engaged in silent immobile *observation* of the skill as was demonstrated by the first group of subjects who had the opportunity to engage in physical practice of the skill with their other hand [EBERHARD, 1963].

TOLMAN's [1932] cognitive map theory was derived from the fact that animals learned the general conformations of mazes independent of movements made when traversing them. Some contemporary theorists espousing the overriding importance of movement in the formation of the intellect seem to refute the ability of humans to think without chaining together movement responses and without depending upon a background of successful motor experiences [KEPHART, 1960].

Proponents of various cognitive theories of learning, however, are more likely to be receptive to the findings emanating from experiments that explore the effects of mental practice upon motor skills. These investigations have usually demonstrated that thinking about a skill, combined with some physical practice, is more helpful than either physical or mental practice alone [CRATTY, 1967].

A recent investigation of mental practice published in the United States illustrates just *when* during the learning process that mental practice should be engaged in. STEBBINS [1968] discovered that mental practice during the initial stages of skill acquisition was as helpful as was physical practice during the same period of time.

Various types of ideo-motor training have been engaged in by European athletes participating in sports with successful results. Team-sport athletes have been taught tactics. The success of slalom skiers has been enhanced by aiding them to better learn the nature of the courses that they are to traverse, while athletes engaging in the triple jump have improved their efforts by mentally rehearsing the movements in the event for a period of months [VANEK and CRATTY, 1969].

A theory of teaching relating cognition to movement experiences has been advanced by MUSKA MOSSTON [1966] at Rutgers University in the United States. I believe that the spectrum of teaching styles advocated by MOSSTON is exerting a positive influence upon physical education and general education, as children are brought from a situation in which they simply respond to the command of the teachers to a stage in which they are encouraged to solve problems through movement.

In a recent monograph some of MOSSTON's ideas have been incorporated into a series of movement tasks through which retarded and immature children may be encouraged to lengthen their attention spans, to recognize geometric patterns and letters, to improve serial memory abilities, to spell and to count. With further research applications of MOSSTON's theory to other types of education, industrial and military situations should be forthcoming [CRATTY, 1969].

Arguments for the fact that all learning is based upon the quality of the early motor responses seen in infants and in children have come from several sources during the past few years. PIAGET [1963] argues for the importance of an initial sensory-motor period which purportedly undergrids later cognitive development. NEWELL C. KEPHART [1960], an American psychologist, has also proposed a perceptual-motor theory which suggests that motor activity contributes in a positive way to the development of perceptual and cognitive processes. PIAGET suggests that early infantile reflexes somehow blend into voluntary movements, whereas KEPHART writes that movement activities aid in the organization of near and distant space.

Others, on the other hand, argue against such a simplistic view of child development. AUSUBEL [1960] and PEIPER [1963] present evidence that most infant reflexes disappear prior to the emergence of voluntary movement attributes. ILLINGWORTH's [1967] schema suggests that as children mature their attribute patterns become more diffuse, making it less likely that various mental, motor and perceptual scores will correlate highly with one another.

MEYERS and DINGMAN [1960], after conducting a survey of factoral studies, find that mental and motor traits reside in different domains by the age of two years, while innumerable studies have demonstrated the inability to predict later intelligence by scores in motor ability tests administered during the first years of life [BAYLEY, 1968; BLOOM, 1964].

Conclusion

The available evidence suggests that 'motor' represents only a facet of the human personality. Lemurs living 23 million years ago evidenced superior manipulative abilities: most animals can run and jump and engage in more spectacular locomotor behaviors than can humans. Only in the human, however, is found the large associational areas constituting about 90% of the neurocortex.

It is perhaps because of this relatively recent appearance of structures which mediate high level thought in humans that they have difficulty at times in co-ordinating thought with motor skill, and why excess pre-occupation with mechanical principles on the part of the performer will sometimes interfere with smooth motor output.

Future directions for research seem endless within this general problem area. Correlative as well as causal studies should prove valuable. It is apparent that, as one increases the complexity of motor tasks presented to groups, increasingly higher correlations will be found between IQ scores and the measures of motor performance collected. The exact manner in which thought provoking movement education programs contribute to the acquisition of what cognitive abilities, however, remains somewhat obscure at the present time.

Further research should also be concerned with how mental rehearsal interacts with tasks of various levels of difficulty. When and how should sports performers and industrial workers be encouraged to think about the execution of skill important to each provide another interesting avenue of study.

It is apparent that man is both a thinking and a moving animal. The exact conditions under which both facets of his personality may be supportive or disruptive of one another, however, still remains somewhat obscure. To ignore the role of cognitive processes in the acquisition of skill seems as naive as attempting to improve reading by walking balance beams.

References

DOMAN-DELACATO treatment of neurologically handicapped children. American Academy of Pediatrics Executive Board Statement, AAP News Letter *16:* 11 (1965).

AUSUBEL, D. P.: A critique of PIAGET's theory of the ontogenesis of motor behavior. J. genet. Psychol. *109:* 119–122 (1966).

BAYLEY, NANCY: Behavioral correlates of mental growth-birth to 36 years. Amer. Psychol. *23:* 1–17 (1968).

BLOOM, B. S.: Stability and change in human characteristics (Wiley, New York 1964).

CORDER, W. D.: Effects of physical education on the intellectual, physical and social development of educable mentally retarded boys. Unpublished special project (George Peabody College, Nashville 1965).

CRATTY, B. J.: Development sequences of perceptual motor tasks. Freeport, Long Island: Educational activities (1967). – Learning and playing: fifty games for children with learning difficulties. Freeport, Long Island. Educational activities

(1967). – Movement behavior and motor learning; 2nd ed. (Lea & Febiger, Philadelphia 1967). – Perception, motion and thought (Peek Publications, Palo Alto 1969). – Active learning (Prentice-Hall, Englewood Cliffs 1971).

EBERHARD, U.: Transfer of training related to finger dexterity. Percept. Motor Skills *17:* 274 (1963).

EGSTROM, G.: The effects of an emphasis on conceptualizing techniques upon the early learning of a gross motor skill; diss. Los Angeles (1961).

HUMPHREY, J.H.: Comparisons of the use of active games and language workbook exercises as learning media in the development of language understandings with third grade children. Percept. Motor Skills *21:* 23–26 (1965).

ILLINGWORTH, R.S.: The development of the infant and young child, normal and abnormal (Livingstone, Edinburgh 1967).

ISMAIL, A.H. and GRUBER, J.J.: Motor aptitude and intellectual performance (Merrill, Columbus 1967).

ISMAIL, A.H. and KIRKENDALL, D.R.: Relationships among three domains of development. Speech presented to the 2nd International Congress of Sport Psychology, Washington, D.C., 1968.

JACOBSON, E.: Anxiety and tension control. A physiologic approach (Lippincott, Philadelphia 1964). – Progressive relaxation; 2nd ed. (University of Chicago Press, Chicago 1938). – Tension in medicine (Thomas, Springfield 1967).

JOHN, E.R.: Mechanisms of memory (Academic Press, New York 1967).

KEPHART, N.C.: The slow learner in the classroom (Merrill, Columbus 1960).

KERSHNER, J.R.: DOMAN-DELACATO's theory of neurological organization applied with retarded children. Except. Children 441–450 (1968).

LA BARRE, W.: The human animal (University of Chicago Press, Chicago 1954).

LAVERY, J.J.: Retention of a skill following training with and without instructions to retain. Percept. Motor Skills *18:* 275–281 (1964).

MACCOBY, E.E.; DOWLEY, E.M., and HAGEN, J.W.: Activity level and intellectual functioning in normal and pre-school children. Child Develop. *36:* 761–769 (1965).

MEYERS, C.E. and DINGMAN, H.F.: The structure of abilities at the pre-school ages: hypothesized domains. Psychol. Bull. *57:* 514–532 (1960).

MILNER, E.: Human neural and behavioral development (Thomas, Springfield 1967).

MOSSTON, M.: Teaching physical education from command to discovery (Merrill, Columbus 1966).

MURPHY, G.: Human potentialities (Basic Books, New York 1958).

OLIVER, J.N.: The effects of physical conditioning exercises and activities on the mental characteristics of educationally sub-normal boys. Brit. J. educ. Psychol. *28:* 155–165 (1958).

OLSON, W.C.: Child development (Heath, Boston 1959).

PEIPER, A.: Cerebral function in infancy and childhood (Consultants Bureau, New York 1963).

PIAGET, J.: The psychology of intelligence (Littlefield Adams, Patterson 1963).

ROBBINS, M.P. and GLASS, G.V.: The DOMAN-DELACATO rationale: a critical analysis; in JEROME HELLMUTH Educational therapy, vol. 2 (Special Child Publications, Seattle 1968).

SOLOMON, A. and PRANGLE, R.: Demonstrations of physical fitness improvement in the EMR. Except. Children 177–181 (1967).

STEBBINS, R. J.: Relationship between the perception and reproduction of certain motor skills. Res. Quart. amer. Ass. Hlth phys. Educ. *39* (1968).

TOMAN, E. C.: Purposeful behavior in animals and men (The Century Co., New York 1932).

VANEK, M and CRATTY, B. J.: Psychology and the superior athlete (MacMillan, New York 1969).

Author's address: Prof. BRYANT J. CRATTY, University of California, Perceptual-Motor Learning Laboratory, Department of Physical Education, *Los Angeles, CA 90024* (USA)

Medicine and Sport, vol. 6: Biomechanics II, pp. 100–108
(Karger, Basel 1971)

Proprioceptive Integration and Motor Learning

JOAN C. WATERLAND

University of Wisconsin, Madison, Wisc.

Fortunately, lack of knowledge of the human nervous system, which ECCLES in 1958 called, 'the most complex machine existing today', does not prevent us from making use of overt and covert manifestations in the operation of the motor learning process. The physiological basis of all motor learning is apparent in the structural and functional organization of the nervous system, and this is not haphazard. Cortical and subcortical integrative centers modulate sensory input and regulate motor output flowing from specific receptors to appropriate effectors. Thus, information regarding muscle length and tension, joint position or movement, head and body position, touch and pressure is constantly being relayed to all parts of the central nervous system, which provides for sensory integration before final motor expression.

In volitional movement the spatio-temporal configurations of motor activity are continuously being regulated and modulated by sensory input, which integrates both the willed and the reflex components of movement. CRITCHLEY [1954] so aptly considered this integration when he said, '. . . all associated muscular activity becomes regulated, though not at the conscious level, so as to form a harmony of movement, of which the prime movers constitute the melody.' The importance of the supportive or reflex components of movement in motor learning is also in agreement with writings of HINES [1943], HESS [1954], and PAILLARD [1960].

The work of the Motor Learning Research Laboratory is primarily concerned with the study of 'natural' movement patterns performed by 'normal' subjects and with the relationships existing between the reflex and willed components of movement. This paper will include aspects of motor learning which relate to: (1) reflex activity underlying movement and (2) the acquisition of volitional movement.

Reflex Activity Underlying Movement

Sir CHARLES SHERRINGTON [1950], the father of modern day neurophysiology, has stated, '. . . reflex action has contributed much to the integration of the individual. It would seem to be reflex action, not mind, which primarily integrates the motor individual.' It is, therefore, important to be aware of the basic reflex patterns which underlie efficient movement, and these include reactions at lower and higher levels of the central nervous system. Receptors which stimulate lower-level reflex actions are found in muscles, tendons, ligaments, joints fascia, and skin. Higher-level balance and equilibrium reactions are important in the regulation of supportive attitudes during the performance of any motor activity. Physiologists have often stressed that volitional movement is dependent not only upon co-ordinated muscular actions but also upon appropriate reflex adjustments.

Limb Synergies

Associated motor activity taking place in muscle groups far removed from actively contracting muscles is noted during reciprocal motions at the wrist joint. Alternate flexion and extension wrist movements have been found to elicit action potential activity in the ipsilateral brachialis muscle during volitional wrist flexion against gravity and excitation in the lateral head of the triceps muscle during active wrist extension. The irradiation of muscular activity observed in the synergistic musculature of the ipsilateral extremity may be attributed to proprioceptive influences from the initially contracting muscles. As early as 1927, COOPER and CREED demonstrated that contraction of the tibialis anterior muscle evoked flexor patterning in the entire ipsilateral limb. The total limb synergy thus elicited from the sensory inputs of actively contracting muscles may be regarded as the reflex or supportive background necessary for volitional movements.

In the radioulnar joints, nature provides an admirable device for varying reflex reactions, and GELLHORN [1953] was the first to study this extensively in relation to proprioceptive inputs. He evoked what he called a 'biceps and a triceps complex' by electrically stimulating appropriate cortical areas in rabbits, cats, and monkeys and subsequently demonstrated that muscle patterning could be changed by varying the radioulnar joint position. In 1948 he demonstrated the same patterning complexes in humans by using voluntary isometric contractions in the upper extremities. Wrist ergographic exercises designed to stress a subject systematically and bring him to the limits of his functional capacity were later used by WATER-

LAND and HELLEBRANDT [1964] to differentiate influences emanating from the prime movers activated volitionally from inputs elicited by modifying the position of the radioulnar joints. A total flexor limb synergy was called forth when wrist flexion was performed against resistance with the radio-ulnar joint in the anatomical or supinated position. Wrist extension performed from the same radioulnar positioning augmented a limb synergy comprised of extensor muscles. The action of supportive musculature was in harmony with that of the prime mover. However, the same wrist movements performed from a position of pronation evoked activity in associated muscles which was in conflict with the volitional act; wrist extension was linked with a flexor limb synergy and wrist flexion with an extensor synergy.

DUCHENNE wrote as early as 1835: 'An isolated muscular action does not exist in Nature.' The study of muscle interdependence has revealed certain predetermined configurations which suggest that the whole concept of isolated muscle activity has little basis in physiologic fact. The interweaving of muscle action, even from distantly located muscles, is demonstrated by the overflow of excitation or irradiation which occurs when resistance is given a particular movement. Similarly, a volitional joint action produces involuntary movement at the neighboring articulations with predictable joint patternings when the muscles involved are placed anatomically to affect more than one articulation, as demonstrated with the radioulnar and elbow joints [OLSON and WATERLAND, 1967]. For example, volitional elbow flexion-extension movements performed from a supinated radioulnar position linked supination with flexion and pronation with extension in the sagittal plane. However, volitional radioulnar supination-pronation in the frontal plane combined supination with elbow extension and pronation with flexion.

Head-Shoulder Linkages

The radioulnar joint has a profound influence on the upper appendage during ergographic exercise and also affects the patterning of its contiguous parts, thus implicating the shoulder girdle, spine, and head. Covert and overt behavioral changes occurred in association with the maintenance of capacity in the face of mounting stress [WATERLAND and HELLEBRANDT, 1964]. A head-shoulder girdle linkage was demonstrated by all subjects, and this appeared to be a physiological constant under the conditions imposed. Some of the stereotypes that were identified are as follow: *First,* unilateral shoulder girdle retraction was always associated with the ipsilateral head turning while protraction of the shoulder girdle was always linked with contralateral head turning. *Second,* cervical spine ventral flexion was associated with shoulder girdle protraction and cervical spine dorsi-flexion was tied with shoulder girdle retraction.

The head-shoulder girdle linkage was in no way affected by substitution of stronger proximal muscle groups for the wrist joint agonist or by discordant proprioceptive inputs. The invariability of the head-shoulder linkage was so impressive that a study of 'non-stressful' volitional upper extremity, shoulder girdle and head movements was undertaken [WATERLAND and MUNSON, 1964]. This investigation showed that the head, shoulder girdle, and glenohumeral joints are mutually dependent; activity in one affected the behaviour of the others, and the linkages observed were then classified as two-way stereotypes.

The movements of a child with central nervous system deficits who exhibited marked tonic neck reflex activity were investigated. It was found that when an experimenter blocked shoulder girdle movement unilaterally, stereotyped tonic neck reactions were disrupted during the time of stabilization. The youngster could turn her head away from the side of shoulder stabilization, with considerable effort, in response to her name being called. Head rotation did not elicit flexion of the upper extremity on the skull side where the shoulder was fixed. Photographs taken during unrestricted head rotation movements definitely revealed shoulder girdle involvement, which is in keeping with head-shoulder linkages. Historically only one researcher has alluded to more proximal body parts in the 'classical' tonic neck reflex literature.

The primitive stereotyped movements characteristic of the tonic neck reflexes in infants or children with central nervous system deficits never become completely submerged in the motor repertory of 'normal' adults. Vestiges of the reflex are later expressed as the head-shoulder linkages. TWITCHELL has stated [1965], '... wholly new and distinct reactions are not added at successively higher levels of the nervous system, but more primitive reactions become modified and elaborated as the stimulus of their response becomes more discriminating.' The mechanism responsible for integration of the head-shoulder linkage observed in adults might be attributed to reflexes arising as a result of stimulation of the cervical joint receptors (C 1, 2, 3) and/or the muscle spindles in the neck and shoulder area.

Postural Compensation in an Amputee

Further evidence of the importance of reflex adjustments for adequate motor behavior is illustrated by the remarkable compensatory ability of a quadrilateral congenital amputee during postural stability tests. The subject of this study had bilateral Chopart amputations of the lower extremities and an above- and a below-elbow amputation of the upper extremities. Normally, the biped position is continually being threatened by gravitational stretches on anti-gravity musculature, and appropriate postural adjustments have to be constantly made in order to maintain stable upright positions. The postural integrity of any quadrilateral amputee would, therefore, appear to be hampered because of seemingly impaired sensory inputs incurred by the amputation. In spite of his handicaps, the subject in this study was effectively able to compensate for his disabilities to such a degree that his postural stability superseded that which is usually observed in normal individuals. HELLEBRANDT [1950] observed that normal man uses 66 % of his functional base during dynamic stance conditions, whereas the subject of this study utilized about 39 % of the base available to him. Therefore, highly developed lower level-reflex actions coupled with higher level balance and equilibrium responses were probably responsible for the great degree of postural stability noted during adverse pathological conditions.

Acquisition of Volitional Movement

PAILLARD [1960] has stated that there are two steps in learning a new motor activity: the *first* requires cortical participation to overcome difficulties encountered and the *second* necessitates automatic movements which are almost completely independent of cerebral control. The transition from cortically directed movements to more subconscious levels occurs during the learning process. HESS [1954] believes that the level of skill achieved will be determined by the degree to which the body can anticipate and support willed movement. Reflexes provide 'dynamic support' upon which willed movement is superimposed. By identifying and studying the reflex components of willed movement in the Motor Learning Research Laboratory, it is hoped that 'key' understandings can be translated into 'key' cues which will then facilitate motor learning.

Muscle Patterning and Skill Achievement

MILANI-COMPARETTE and GIDONI [1967] have defined muscle patterning to be '. . . selected activation of certain muscle groups in such a combination of lengths that the result is a definite movement. The pattern is thus a reflection of a spatio-temporal configuration of nervous excitation and inhibition of the motor-neuron pool and the pattern at the spinal level is a resultant pattern integrated from other patterns existing at higher levels of the central nervous system . . . Some of the motor patterns are recognizable as the stock of reflexes and primary automatisms. Others are added by learning. Together they constitute the repertory of patterns upon which the motor system can draw for its performance.'

PAILLARD [1960] divided the overall common characteristics of muscle patterning into two components: (1) the spatial disruption of the activity of the various muscles depending on their participation in the specific movement; (2) the temporal succession, the timing and sequence of activation of muscles. The neurophysiological bases of patterning must also be defined in terms of co-ordination [PAILLARD, 1960]. SHERRINGTON [1947] stated that by volitional practice, simple movements can be combined into new movement sequences which become habitual in timing since their execution no longer requires concentration of attention. In the initial phase of motor learning, a generalized tension exists in the musculature which may be caused by an excessive amount of cortication; however, through repetition and practice the movement becomes more independent of cortical control [PAILLARD, 1960]. Skill is acquired through practice which is evaluated and is partly due to the refinement of muscle co-ordination.

The common characteristics of efficient muscle activity include delicacies of adjustment, economy of execution, and accuracy of achievement. The ability to control skeletal muscles toward a given end is dependent upon the co-ordinated action of co-operative muscles and upon complex postural adjustments. PAILLARD [1960] suggested that an essential component to efficient movement is the ability to restrict joint actions to those necessary to make the movement. STEINHAUS [1964] proposed that skill depends as much on suppressing unproductive muscle action as it does on effecting the correct timing of maximal force in selected muscle groups. Skillful motion always involves the least expenditure of effort [STEINDLER, 1935].

Head-Shoulder Linkages and Skill Achievement

Head-shoulder linkages may be related to motor learning since some movements seem to rely on suppression of the innate pattern associates while others depend upon the preservation of the natural linkages. Fundamental motor activities taught to young children or to beginners of any age are probably characterized by the retention of natural and strong head-shoulder two-way linkages, while complex activities may be more dependent on a volitional suppression of these natural pattern associates.

Cinematographic recordings of a 2½-year-old child throwing a ball suggest that the young do not have the capacity to disturb volitionally the natural yoke of the shoulder girdle to the head. The child stands with feet in a side-stride position, with the shoulders and head facing the direction of the intended throw. The activity is mostly restricted to movement in the glenohumeral joint. Perhaps this is the natural developmental pattern which must be later modified to attain a more forceful throw.

One of the most common teaching cues for a beginning tennis player is 'side to the net'. A novice has a natural tendency to face the net and attempts to hit the ball from this position without interfering with innate head-shoulder linkages. However, the golf swing demands that one separate head mobility from shoulder girdle movement. A skillful golfer learns to rotate the hips, trunk, and shoulders between two stationary privots, the feet and the head. A skilled bowler has developed by cortical control the ability to let the arm swing against a stabilized shoulder girdle. The latter enhances the accuracy of the throwing arm, and the associated head movements are negligible.

Violin playing requires an unnatural head-shoulder positioning. BARLOW [1954] reported that it is not uncommon for violinists to complain of shoulder and arm muscle spasms. A violinist, a patient of BARLOW's, '. . . after playing for a while begins to pull his head over to the right . . . [and] he cannot prevent it when he becomes engrossed in his playing . . . even after his position has been corrected.' He assumes a natural head-shoulder stereotype instead of the one dictated.

Neurophysiological mechanisms, which include proprioceptive and exteroceptive influences, may be considered as a phase of a continuum; another phase may be represented by psychological factors. The importance of one should not be overemphasized or be considered adequate to explain motor learning since both are mutually dependent. The balance between the physiological and psychological components must also be in harmony with physical and biomechanical factors since they are also linked in the continuum of dependency. DAVID KRECH [1961], who said, 'understanding atomic physics is child's play compared with the understanding of child's play', adroitly summarizes the complexities inherent in understanding movement and thus motor learning.

References

BARLOW, W.: Posture and the resting state. Ann. phys. Med. *2:* 113–122 (1954).

COOPER, S. and CREED, R. S.: Reflex effects of active muscular contraction. J. Physiol., Lond. *62:* 273–279 (1927).

CRITCHLEY, M.: Discussion on volitional movement. Proc. roy. Soc. Med. *47:* 593–601 (1954).

DUCHENNE, G. B.: Physiology of motion: translated and edited by KAPLAN (Saunders, Philadelphia 1959).

ECCLES, J. C.: Physiology of imagination. Sci. Amer. Sept. (1958).

GELLHORN, E.: Physiological foundations of neurology and psychiatry (The Univ. of Minnesota Press, Minneapolis 1953).

HELLEBRANDT, F. A.: Influence of lower extremity amputation on stance mechanics. J. amer. med. Ass. *142:* 1353–1356 (1950).

HESS, W. R.: Diencephalon-automatic and extrapyramidal functions (Grune & Stratton, New York 1954).

HINES, M.: Control of movements by the cerebral cortex in primates. Biol. Rev. *18:* 1–31 (1943).

KRESH, D.: Control of the mind. A symposium (McGraw-Hill, New York 1961).

MILANI-COMPARETTI and GIDONE, E. A.: Pattern analysis of motor development and its disorders. Develop. Med. Child Neurol. *9:* 625–630 (1967).

OLSON, J. and WATERLAND, J. C.: Behavior of independent joints served in part by muscles common to both: elbow and radioulnar joints. Percept. Motor Skills *24:* 339–349 (1967).

PAILLARD, J.: The patterning of skilled movements; in FIELD, MAGOUN and HALL Handbook of physiology (American Physiological Society, Washington, 1960).

SHERRINGTON, C. S.: The integrative action of the nervous system; 2nd ed. (Yale Univ. Press, New Haven 1947).

SHERRINGTON, C. S.: Man and his nature (Doubleday, New York 1950).

STEINDLER, A.: Mechanics of normal and pathological locomotions in man (Thomas, Springfield 1935).

STEINHAUS, A. H.: Toward an understanding of health and physical education (Brown, Dubuque 1963).

TWITCHELL, T.E.: Attitudinal reflexes; in The child with central nervous system deficit (US Government Printing Office, Washington 1965).

WATERLAND, J.C. and HELLEBRANDT, F.A.: Involuntary patterning associated with willed movement performed against progressively increasing resistance. Amer. J. phys. Med. *43:* 13–30 (1964).

WATERLAND, J.C. and MUNSON, N.: Reflex association of head and shoulder girdle in nonstressful movements of man. Amer. J. phys. Med. *43:* 98–108 (1964).

Author's address: Dr. JOAN WATERLAND, Motor-Learning Research Laboratory, 321 Lathrop Hall, University of Wisconsin, *Madison, WI 53706* (USA)

Medicine and Sport, vol. 6: Biomechanics II, pp. 109–113
(Karger, Basel 1971)

The Formation of a Competitive Motor Response by Operant Conditioning

CLAUDINE Y. JEANRENAUD, ANTHONY G. LINFORD, M. VIRGINIA FRYE
University of Illinois, Champaign, Ill.

Introduction

Operant conditioning, or behavior modification, represents a technology that is receiving much emphasis in the field of psychology. The operant technologist would, basically, maintain that the frequency of a behavior can be modified in accordance with the kind of reinforcement that follows the behavior. While operant technology has been widely used as a research technique by clinical and developmental psychologists, little attention has been paid to its possible value as a tool for producing, and modifying, gross motor responses. It would seem, however, to have potential for application in the areas of physical medicine and rehabilitation, particularly with mentally retarded patients.

Review of Literature

Young retarded children, characteristically, are difficult to motivate to expend high levels of energy [Amer. Ass. Hlth, 1968]. As the factor of motivation enters most equations for the calculation of mechanical output by the human organism, this represents a major bio-mechanical problem.

STEIN and PANGLE [1966] have alluded to the fact that many retardates have to be taught basic motor skills that the non-retarded acquire by incidental learning. Studies on the use of behavior modification as a technology to produce this learning are few, but encouraging. HARRIS *et al.* [1964] were able to modify regressed crawling behaviors in a kindergarten child while JOHNSON *et al.* [1966] improved the activity rate of a young child in a playground setting. The Down's Syndrome (mongoloid) child is prone to upper respiratory tract infections [TREADGOLD and SODDY, 1963] and the need for developing its various physical fitness parameters has been stressed. SECHREST [1964] modified a bicycle ergometer to dispense reinforcers after a certain number of revolutions of the pedals. He was able to increase both the duration and severity of exercise, by trainable Down's Syndrome retardates, by this behavioral procedure. It is doubtful, however, if any of his subjects attained sufficiently high levels of energy expenditure to produce a training effect. On the other hand, this was achieved with similar subjects by LINFORD and DUTHIE [1968]. They applied operant techniques to teach an endurance circuit and used post-exercise heart rate as an indicant of energy expenditure.

In so far as competitive behavior necessitates the interaction of two factors: (a) the desire on the part of at least two people for the same goal, and (b) an attitude towards others that are seeking the same goal [GROSSACK, 1954], it can, thus, be considered a learned behavior. The positive motivational effects of competition upon the performance of high endurance motor activities has been frequently noted by physical educators. In two studies [WOLF, 1938; MCMANUS, 1967] in which competition was used as an incentive, more persistence of behavior was produced by the competition group than by groups motivated by praise, reproof, or interest intrinsic to the task. On the other hand, where failure is frequently experienced, competitive discouragement, or depressed performance in the competitive situation has been observed [LEIBOWITZ, 1966]. BAILER and CROMWELL [1966] studied this phenomenon and concluded that moderate failure could have motivating effects. Reanalysis by the present authors of the data presented in an older study [WINSLOW, 1940] confirmed this. All rats that were reinforced on 50% or more of their competitive trials increased in running speed.

This study was designed to investigate if operant conditioning techniques could be used to produce a competitive response in young trainable retarded Down's Syndrome children. A competitive response was operationally defined as running faster in a group situation than when alone.

Method

The subjects were 10 Down's Syndrome children aged 5–7 years, having IQ's in the 30–50 range. The experimental procedures were divided into two parts. First the subjects were taught to perform the task correctly. The task involved running a distance of 25 yards down a wide corridor, touching an end wall and running back to the experimenter. During this phase all subjects had an equal number of trials alone, and in groups. Continuous reinforcement was given for correct completion of the task (irrespective of order of finishing in the grouped trials). When all subjects could perform the task correctly, the main experimental treatment began. Baseline times were recorded for all subjects running both individually and in groups of two. The subjects were then given 8 trials per day, with 2 subjects per group, 5 days per week for a period of 23 days. Trials were structured, by means of handicaps, so that every child won 4 trials per day. Groupings were different from trial to trial and from day to day. The winner of each trial was given social reinforcement in the form of praise and physical contact; the loser of each trial was ignored. Thus, the subjects had to learn that differential reinforcement was given for winning. On the 24th day all subjects were again tested in groups of two and alone. Time in both pre- and posttest situations was recorded to the nearest 0.5 second. The order of testing (group or individual) was randomly assigned. It was hypothesized that, following treatment, subjects would run faster in the group (competition) condition than when alone (non-competition).

An analysis of variance for a single factor experiment involving repeated measures was used to analyze the data (table I). A significant ($p < 0.01$) treatment effect was detected.

Differences between pairs of means were inspected by use of the Newman-Keuls test. This revealed that performance in the posttest competition condition was significantly superior ($p < 0.05$) to that under any other condition. There were no significant differences ($p > 0.05$) between any other pair of means (table II).

Table I. Analysis of variance for repeated measures

Source	S.S.	df	M.S.	F
Between people	1881.0	9	209	
Within persons	1233.6	30	41.12	
Treatments	658.4	3	219.5	10.035[1]
Residual	575.1	27	21.3	
Total	3115.3	39		

1 $p < 0.01$

Table II. Test on differences between pairs of means

	Posttest		Pretest	
	Comp.	Alone	Alone	Comp.
Total	203.0	274.0	293.0	309.5
203.0	—	71.0[1]	90.0[1]	106.5[1]
274.0		—	19.0	35.5
293.0			—	16.5
309.5				—
$q_{.95\ (r,10)}$		4.48	5.27	5.77
$\sqrt{n MS_{res.}}\ q_{.95\ (r,10)}$		65.408	77.99	84.24

1 $p < 0.05$

The lack of significant difference between the pretest means indicates that, prior to the experimental manipulation, competition was not a positive motivating factor in performance. This could have been the result of competitive discouragement because of a history of failure that is typical of the retarded child, or, alternatively, the children had not learned to compete. Differences in the posttest means, in favor of the competitive situation, were in the predicted direction. As an alternative hypothesis to explain this change the phenomenon of social facilitation could be considered. WINSLOW [1940] observed that rats tended to run faster through a maze when in groups. Similarly, HARLOW [1932] and HARLOW and YUDIN [1933] observed that both rats and monkeys ate more in groups than when they were fed individually. Social facilitation, however, would not appear to explain the differences found in this study. Firstly, all the subjects were familiar with each other, having lived together for a period of 9 months, whereas, in the animal studies the subjects were kept in separate cages. Secondly, if social facilitation had occurred, one would have expected the pretest competitive times to have been faster than the individual ones. This was not the case. It can, therefore, be reasonably assumed that the significant superiority of the posttest competitive response was a direct result of the behavior modification techniques which were applied.

This study indicates that young, mentally retarded children can, by the appropriate operant conditioning, be induced to produce superior running performance in a competitive situation. This learned competitive behavior can contribute to superior bio-mechanical performance in high energy cost situations by increasing the potency of the motivational function which is part of the energy output equation. The possibility that increased energy output could reduce the incidence of upper respiratory tract infections in Down's Syndrome children may merit further investigation.

References

American Association for Health, Physical Education and Recreation, Washington: Physical activities for the mentally retarded (1968).

BAILER, I. and CROMWELL, F.: Failure as motivation with mentally retarded children. Amer. J. ment. Defic. *69:* 680–684 (1965).

GROSSACK, M.H.: Some effects of cooperation and competition on small group behavior. J. abnorm. soc. Psychol. *49:* 341–348 (1954).

HARRIS, F.R., and others: Effects of positive social reinforcement on regressed crawling of a nursery school child. J. educ. Psychol. *55:* 35–41 (1964).

HARLOW, H.F.: Social facilitation of feeding of the albino rat. J. gen. Psychol. *41:* 211–221 (1932).

HARLOW, H.F. and YUDIN, H.D.: Social behavior of primates. Social facilitation of feeding in the monkey and its relation to attitudes of ascendance and submission. J. comp. Psychol. *16:* 171–185 (1933).

JOHNSON, M.K., and others: An application of reinforcement principles to the development of motor skills of a young child. Child Develop. *37:* 379–387 (1966).

LEIBOWITZ, L.F.: The motivation effect of values symbols and competition upon problem solving behavior in children. J. gen. Psychol. *108:* 327–332 (1966).

LINFORD, A.G. and DUTHIE, J.H.: The use of operant technology to induce sustained exertion in young trainable Down's Syndrome children. 2nd int. Congr. Sport Psychol., Washington, D.C., November, 1968.

MCMANUS, D.L.: Marble-sorting persistence in mixed verbal-incentive and performance-level pairings. Amer. J. ment. Defic. *71:* 811–817 (1967).

SECHREST, L.: Exercise as an operant response of retarded children. Progr. Report NIMH Grant No. M5749 (1964).

STEIN, J.U. and PANGLE, R.: What research says about psychomotor functions of the retarded. J. Hlth phys. Educ. Recreat. *37:* 36–38 (1966).

TREADGOLD, R.S. and SODDY, K.: Textbook of mental deficiency-subnormality; 10th ed. (Boilliere, Tindall & Cox, London 1963).

WINSLOW, C.N.: A study of experimentally induced competitive behavior in the white rat. Comp. Psychol. Monogr. *15:* 1–35 (1940).

WOLF, T.H.: Persisting behavior of kindergarten children (University of Minnesota Press, Minneapolis 1938).

Authors' address: Dr. CLAUDINE Y. JEANRENAUD, Dr. A.G. LINFORD, and M. VIRGINIA FRYE, University of Illinois, Room 97, *Champaign, IL 61820* (USA)

Medicine and Sport, vol. 6: Biomechanics II, pp. 114–116
(Karger, Basel 1971)

On the Problem of Dynamic and Static Periods of Work Activity of Apprentices in Relation to Training

M. VANĚČKOVÁ and L. KOMÁREK

Institute of Hygiene, Prague

Previous work [KOMÁREK, 1968] reported at Zurich in 1967 shows that the rhythm of paced work is individual, and forcing a certain pace of movement upon a person causes a negative effect upon the formation of the working habit in apprentices. We verified this in a group of students in a model experiment and proved that a low pace set at the start of the experiment increases during the period of training. During the experiment the natural pace decreased, so that after 10 days both paces of the motions during working were practically identical (fig. 1). At the start of the study a higher pulse rate per motion cycle was observed under the working condition of set pace which tends to show a greater effort by which the work was carried out. In similar work it actually tends to show a greater effort due to static strain [VANĚČKOVÁ, 1967] (fig. 2). In view of the fact that the set pace was lower than the optimum one at adolescent age, it may be assumed that for its continuance certain mechanisms for breaking the work must have interfered. This hypothesis leads to the consideration that a mutual relation between dynamic and static periods of work activity exists. We pay attention to the postural muscles and the operational muscles themselves as well as to vegetative functions.

In a series of preliminary experiments we proved for example that the pattern of the electromyograph activity in both muscle groups studied is similar and that the significant deviation in the sense of enlargement during the last 30% of the 10-minute experiment agrees with the results obtained by GUMENER [1967], who explains this phenomenon by the nodal point of autoregulative mechanisms of functions. Further experiments carried out in co-operation with GUMENER confirmed that rhythmical fluctuation of IEA in muscles of workers during dynamic and static work as well as

1, 2

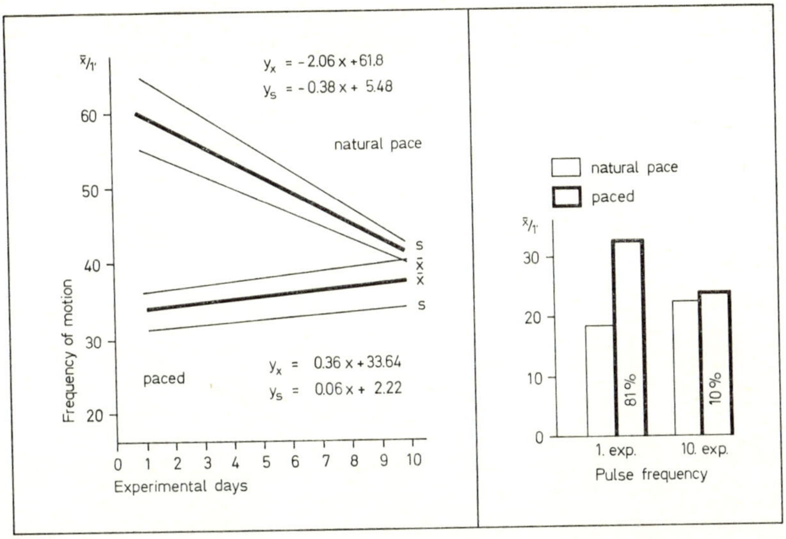

Fig. 1. Trend of the frequency of the working motion in paced and unpaced work (model-filing students).

Fig. 2. Number of pulses over ten motion cycles.

pulse frequency is synchronous in the case of optimal and trained activity (fig. 3).

It appears from our results obtained so far that there is coherence in the activity of the dynamic and the static characters. During training, dynamic activity is characterised by economising and co-ordinating muscle activity, rhythmisation and accuracy of movements and co-ordination of somatic functions with vegetative functions. Static activity possesses a similar character of training although at an absolutely different level. Dynamic and static activity is directed from centres which have their own frequency of operations. Mutual synchronisation of these operations and their synchronisation with vegetative functions should manifest optimalisation of the function of the organism, i.e. manifestation of training from the point of view of regulating physiological functions. On the basis of this working hypothesis we try to solve the problem, in which the main target is to understand the mechanism by which the physiological functions in adolescent organism at various activities are regulated, in order to afford establishment of conditions at which static effort would be decreased to an optimum and learning of the working task would be as favourable as possible.

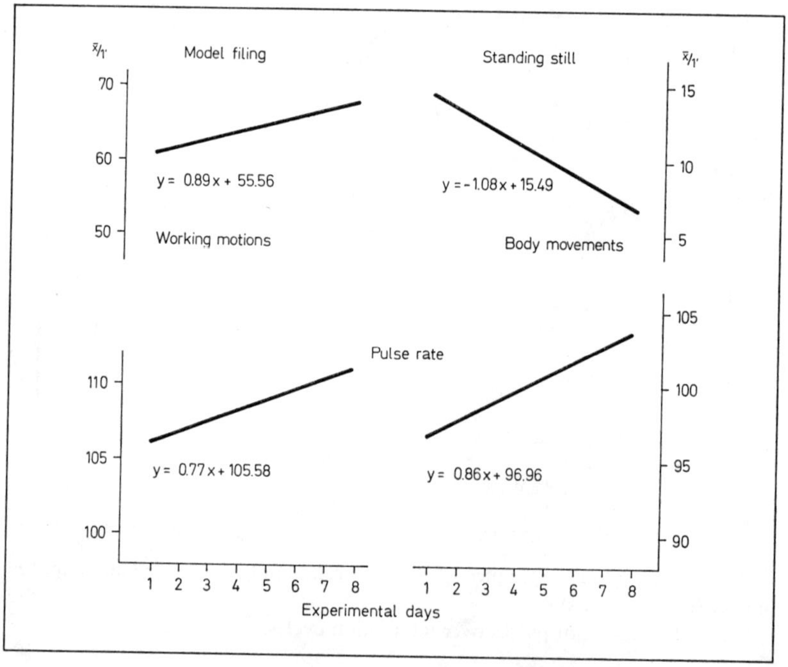

Fig. 3. Trends of the mean values of the number of motions during work, body movements and pulse rate during model-filing and standing still.

References

GUMENER, P. J.: Charakteristika vlijanija fizičeskoj nagruzki na organizm skolnikov. Medicina, Moskva 245 (1967).

KOMÁREK, L.: Biomechanics of training of simple rhythmic efforts. Biomechanics, I. 1st int. Seminar, Zürich 1967, pp. 209–212 (Karger, Basel/New York 1968).

VANĚČKOVÁ, M.: Chances in the organism of apprentices who engage in sports and those who do not through the influence of static fatigue. The II. International Congress about the Physical Fitness of Youth, Prague 1966, pp. 205–209 (Olympia, Praha 1967).

Authors' address: Dr. M. VANĚČKOVÁ and Dr. L. KOMÁREK, Institute of Hygiene, Srobarova 48, *Prague 10* (Czechoslovakia)

Medicine and Sport, vol. 6: Biomechanics II, pp. 117–123
(Karger, Basel 1971)

An Attempt to Examine the Effects of Different Types of Training in Movement of the Arm[1]

H. C. G. Kemper

Laboratory for Psychophysiology and Coronel Laboratory for Occupational Medicine, University of Amsterdam, Amsterdam

I. Introduction

There is no doubt about the importance of training in physical education and sport. Systematic training influences the musculature and consequently the movement, being the result of muscle actions on parts of the skeleton [Nöcker, 1966; Simkin, 1960]. It is very difficult to compare certain aspects of movement as they appear in branches of sport, because each of them has its own functional pattern. This report presents a method which enables us to measure a number of characteristics of movement. The method seems to enable us to record, with objective means, differences due to training or other factors.

II. Experimental Methods

In our experiments on arm movement an apparatus was used which we call *reactograph,* by means of which it became possible to record continuously simple movements of the arm [Baart, *et al.*, 1964; Burger, 1964; Kemper, 1968]. To achieve our aim, we asked our subjects to draw rapidly a line on a moving paper with a ballpoint pen held in a straight groove of the reactograph (fig. 1). The groove is perpendicular to the direction of the movement of the paper, which moves at a constant speed.

Four types of arm movements were chosen as standards, viz., one targetless movement and three to a target indicated by two lines preprinted on the moving paper. When the subject perceived a visual starting signal, he had to perform the movement as fast as he could. The movement had to be stopped between the two target lines.

1 This work was supported by the Netherlands Sports Federation, The Hague, Netherlands.

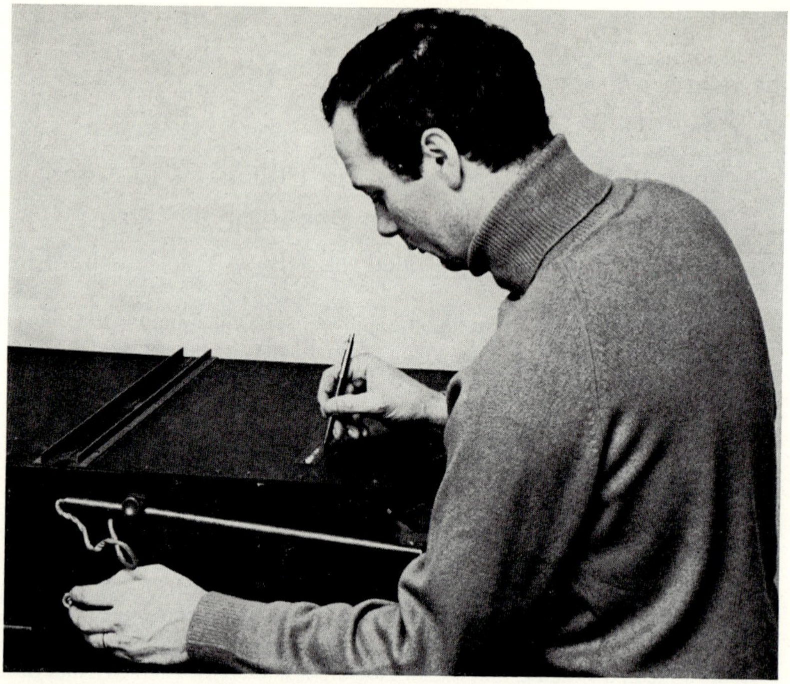

Fig. 1. Reactograph with subject in action

The targets were located at a distance of 30, 12 and 3 cm, respectively, from the starting point and a fourth movement, undirected without a target. Figure 2 shows at the top a series of undirected movements; in the middle and at the bottom two series of directed ones, the target of which was at a distance of 30 and 12 cm, respectively, from the starting point. The results of the movements in relation to time are shown in figure 2.

From these recorded arm movements we tried to determine some aspects according to the criteria which are shown in the diagram of figure 3.

Because the paper moves continuously, the movement of the hand-arm system produces with the aid of the ball point pen a curve on the paper in which the horizontal axis indicates the movement and the vertical axis the time.

The distance from the beginning of the black signal to the starting point of the movement gives the reaction time (RT). The maximum speed (V_{max}) is graphically determined from the displacement-time curve.

The deviations in arriving at the targets have been measured by two scoring methods. One of them, the deviation index (DI), gives general information on the spatial deviation of the inaccuracy of the end of the movement and the deviation time (DT) gives information on the time necessary to correct the deviation.

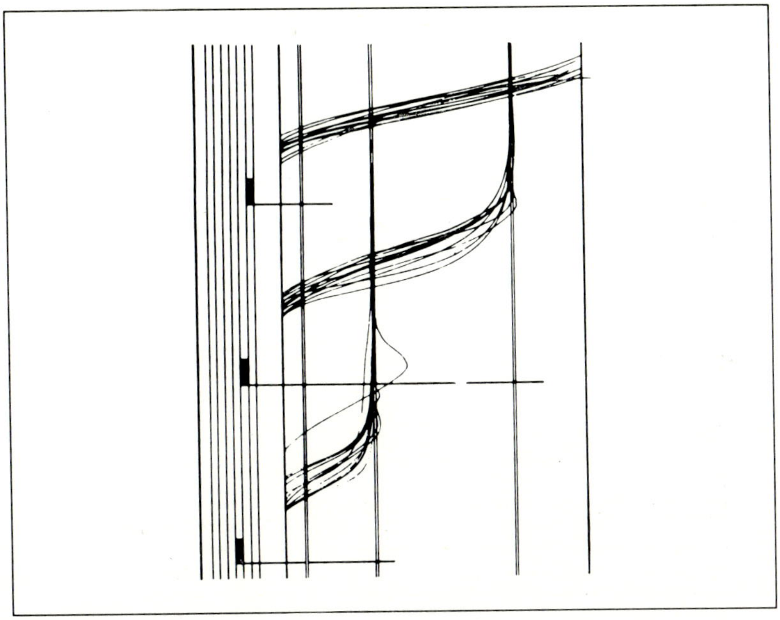

Fig. 2. Reactograph paper with curves of directed and undirected movements

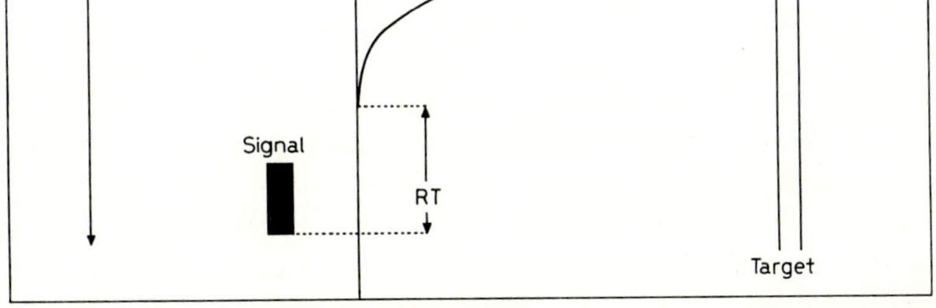

Fig. 3. Diagram of the criteria of measurement

III. Results

The experiment was conducted with 18 top-class athletes and 35 untrained students, all males, as indicated in table I. The top-class athletes had been selected on the performances in their respective fields. The groups consisted of: 6 javelin-throwers, as examples of the training in speed (age 24); 6 weight-lifters, as examples of the training in strength (age 27); 6 table-tennis players, as examples of the training in reaction, anticipation and accuracy (age 22).

Table I. The test groups

Top-class athletes		Untrained undergraduates	
n		n	
6	Javelin-throwers	7	Trained for speed
6	Weight-lifters	7	Trained for strength
6	Table-tennis players	7	Trained for anticipatory action
		7	Trained for endurance
		7	Not trained (for checking purposes)

The progress of the training of each group of athletes was studied over a period of 12 months. All subjects of the group had been trained as uniformly as possible.

On the ground of an inquiry and a medical test the 35 untrained students were selected out of a group of 250. The selection requirement were that during the two years previous to the experiment no form of active sport had been practised. They were divided into 5 groups of 7 persons each, all having about the same age (22–23). In the laboratory 4 of these groups were trained on 5 consecutive days (Monday to Friday), during 6 consecutive weeks: 7 undergraduates trained in arm speed by means of elastic line and punchball; 7 undergraduates trained in arm strength by means of dumbbells and expander; 7 undergraduates trained in anticipatory action by means of a special tableau with steel balls; 7 undergraduates trained in endurance by means of a bicycle ergometer. For purposes of comparison, a fifth group of 7 undergraduates were not trained during this period.

In order to obtain some insight into the course of the achievements, training tests were taken previous to, during, and in many cases 6 weeks after the end of the training. On the basis of these measurements a specific training effect due to the laboratory training could be found. The students trained in strength, for instance, became stronger. Regarding the students who were trained in endurance, the same external load caused a less functional load, measured in lower heart rate and systolic blood pressure.

In order to facilitate the comparison of the effects of the training on the top-class athletes and the undergraduates, all the subjects were tested with the aid of the reactograph (table II). The circumstances of the successive trials to which all the groups of top-class athletes and undergraduates were subjected were:

Table II. The moments of time at which the groups were tested with the reactograph

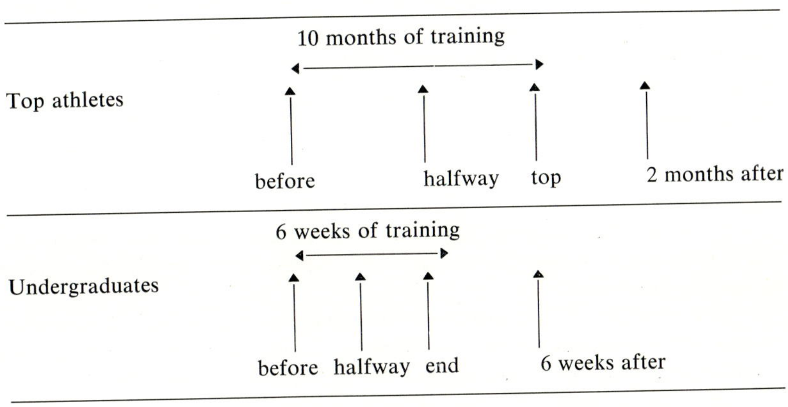

Athletes

Session 1: before the start of the 12-month training
Session 2: halfway the training period
Session 3: at the culminating point in the training period (top)
Session 4: after the training (12 months after session 1).

Undergraduates

Session 1: before the first week of training
Session 2: during the fourth week of training
Session 3: during the first week following the sixth (last) week of the training
Session 4: during the sixth week after the (last week of the) training.

The differences in the data, derived from the curves, produced by the subjects on the reactograph, were statistically tested at a 5-percent significance level. The reliability of the measurements (interjudge correlation) was checked at the same time.

IV. Conclusions

The group of javelin-throwers showed the highest maximum speed at all points of time during the reactograph tests and in each of the four arm movements.

The group of table-tennis players, on the other hand, were most accurate in arriving at each of the three targets, at least as far as indicated by deviation time and deviation index. With respect to these criteria the weight-lifters took up an intermediate position between the other two groups of top-class athletes.

The conclusion was that:

(1) with respect to the groups of top-class athletes trained in various ways the reactograph method might show differences in the performance of movements specific to their respective sports;

(2) the test results give no indication whatever on whether heavy-weight training – as in the case of weight-lifters – causes a decrease in the speed of contraction of the muscles concerned;

(3) with respect to the groups of undergraduates trained in different ways, including the comparison group, no differences could be proved in the pattern of movement, recorded by the reactograph after the 6-week training period.

From the displacement-time curves it was possible to observe two aspects: a latency or reaction time (the lapse of time between the presentation of the starting signal and the beginning of the movement) and a movement time (the duration of the complete movement). Both components comprise integral elements of movement in sport [LOTTER, 1960]. As the rectilinear movements on the reactograph were performed as a reaction on a visual signal, we were able to measure the reaction times as well;

(4) though in the literature results of similar investigations are published which for trained persons show reaction times shorter than those obtained by untrained persons of the same age and sex [WILKINSON, 1958], we have not been able to prove statistically a difference between the group of top athletes and the group of untrained undergraduates.

References

Baart, J.; Ettema, J.H. et Kalsbeek, J.W.H.: L'allongement du temps de réaction en fonction de la complexité du programme à exécuter. Soc. d'Ergonomie de langue française, actes du 2e congrès Bruxelles, 12/13 juin 1964.

Burger, G.C.E.: Permissible load and optimal adaptation. The Society's Lecture Ergonomics Res. Soc. Ergonomics, vol. 7, pp. 397–417 (1964).

Kemper, H.C.G.: Rechtlijnige geleide armbeweging; Diss. (NSF Publication No. 36, The Hague 1968).

Lotter, W.S.: Interrelationships among reaction times and speeds of movement in different limbs. Res. quart. *31:* 147–155 (1960).

Nöcker, J.: Die biologischen Grundlagen der Leistungssteigerung durch Training (Verlag Karl Hoffmann, Schondorf bei Stuttgart 1966).

Simkin, N.W.: Physiologische Charakteristik von Kraft, Schnelligkeit und Ausdauer (Sportverlag, Berlin 1960).

Wilkinson, J.J.: A study of reaction time measures to a kinesthetic and visual stimulus for selected groups of athletes and non-athletes. Proc. College Phys. Educ. *62:* 158–161 (1958).

Author's address: Dr. H.C.G. Kemper, Jan Swammerdam Instituut, le Const. Huygensstraat 20, *Amsterdam* (Netherlands)

Medicine and Sport, vol. 6: Biomechanics II, pp. 124–128
(Karger, Basel 1971)

Effect of Training on the Maximum Isometric Strength in Forearm Flexors

J. NEUMAN and I. KUTA

Physical Education Research Institute, Prague

Values of the maximum isometric strength are among the important factors in the research of muscle activity and mechanical properties of human muscles. They are often used in basic as well as in applied biomechanical research [HILL, 1940; NELSON *et al.,* 1965; WARTENWEILER, 1967; NEUMAN *et al.,* 1968; JOACHIMSTHALER, 1968, and others].

The maximum value of isometric strength (P_0) is an important factor of HILL's characteristic equation expressing the relationship between strength and speed of muscle contraction [HILL, 1940].

The maximum isometric strength measurements depend partly on, among other things, the precise definition of the conditions under which it is possible to reach the maximum values. In this respect we carried out experiments to find answers to the following questions:

With what arrangement could the maximum values of strength be reached?

Is there any difference between the maximum isometric strength in sportsmen and in untrained subjects?

What is the influence of selected somatometric indicators on the value of maximum isometric strength?

Methods

Under standard conditions (fig. 1) the maximum isometric strength of all persons of a group of 15 untrained men (averages: age 20.5 years, weight 73.0 kg, height 173.5 cm) was measured. These results were compared with the results obtained from a group of 17 sportsmen with a high level of physical efficiency (averages: age 21.8 years, weight 75.2 kg, height 177.5 cm). The sport activity of these subjects can

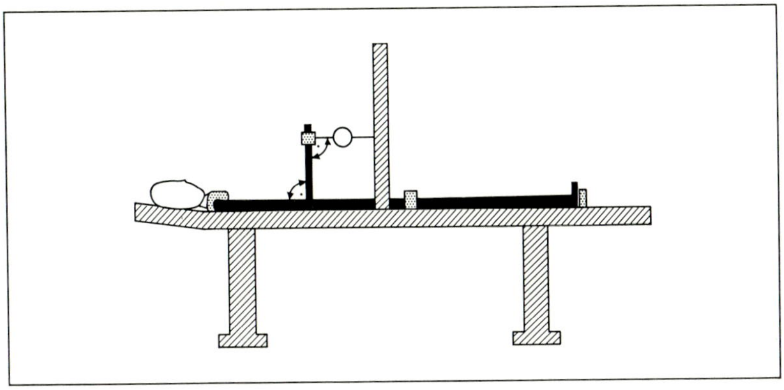

Fig. 1. Position of subject and tensiometer for measuring isometric muscle strength.

be characterised by strength and endurance of work. All subjects were asked to exert maximum isometric force, which was repeated 30 times at maximum voluntary effort. Between every two attempts a relaxation period of 30 sec was given. The measurements were repeated with both groups after an average of 20 days.

Results

The average of the maximum values of muscle strength obtained from sportsmen was 39.5 kgf after the first session and 39.0 kgf after the second. The untrained subjects exerted an average value of 33.7 kgf during the first session and of 34.7 kgf during the second.

The untrained men reached the maximum values of muscle strength between the 6th and 7th trials during the first session and between the 4th and 5th trials during the second session. The sportsmen reached the maximum values of muscle strength between the 3rd and 4th trials during the first session, and after the 4th trial during the second session. In the group of the untrained subjects only 5 men out of the total group of 15 (33.0 %) reached the maximum values of muscle strength after the first three trials during the first session, and only 8 out of 15 men (53.3 %) at the first three trials during the second session. In the group of the sportsmen it was 13 out of 17 (76.5 %) after the first three trials of the first session and 10 out of 17 (58.8 %) of the second session. After the first 5 trials of the first session 10 out of the untrained group of 15 persons (66.7 %) reached the maximum values, and 11 out of this group (73.3 %) after the first 5 trials of the second session. In the group of the sportsmen 15 men out of 17 (88.2 %) reached the maximum values after the first 5 trials of the first as well as of the second session.

Comparing the mean maximum value of force calculated from the first three trials with those of the first five trials it turned out that the mean value of the first three trials always showed to be the highest value (fig. 2). There was one exception: the first trial of the group of the untrained subjects showed that the maximum mean value was reached after three trials of the second session.

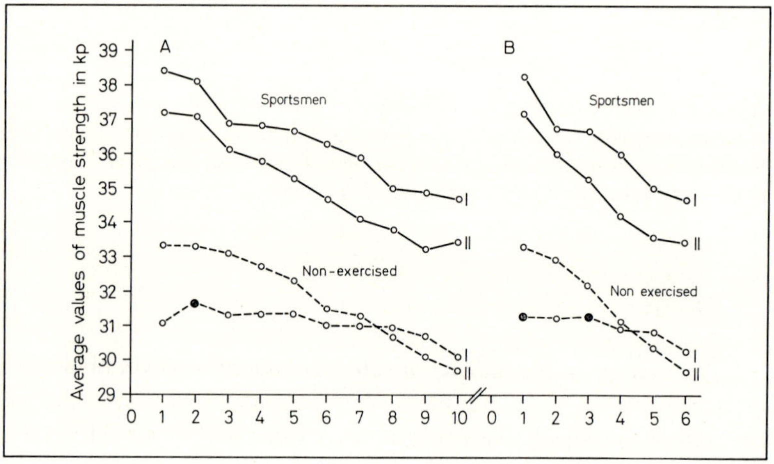

Fig. 2. A Series of three repeated trials; *B* series of five repeated trials.

SPEARMAN's rank correlation coefficient was used for the evaluation of the relationships between isometric strength and the somatic indicators.

TORNVALL [1963] and others have shown that muscle strength is related closely to body weight. In order to exclude the influence of different body weight we calculated the muscle strength per unit of body weight and correlated it with the selected somatic indicators (table I).

Only in the group of sportsmen was a significant relationship between the values of relative muscle strength and height or length of the forearm found, which means that the sportsmen with shorter levers have relatively greater strength than the sportsmen with longer levers. In the group of the untrained subjects this relationship does not exist. We suppose that the formation of these relationships of sportsmen has been due for instance to long and intensive training, which changed the ability to make complete use of the total capacity of muscles.

Table I. Muscle strength/weight ratio, related to untrained subjects and sportsmen

Indicator	Muscle strength/weight ratio			
	Untrained		Sportsmen	
	1st measuring	2nd measuring	1st measuring	2nd measuring
Height	–0.203	–0.329	*–0.547*	*–0.595*
Upper arm length	–0.321	–0.468	–0.339	–0.289
Forearm length	–0.055	–0.159	*–0.494*	–0.476
Arm circ. in contraction	0.190	–0.058	–0.058	–0.027
Arm circ. in relaxation	0.010	0.213	0.014	0.087

Conclusion

(1) The untrained men reached the mean maximum values of muscle strength between the 4th and 7th trials, the sportsmen between the 3rd and 4th trials.

(2) The assumption that all subjects reach the maximum values of isometric strength in the first 3 or 5 trials could not be verified. 11.8 % of the sportsmen and 30.0 % of the untrained subjects who took part in both experiments did not reach their final maximum values after the first five trials.

(3) It may be concluded that sportsmen have the capability to develop their maximum muscle strength under isometric conditions in a smaller number of trials than untrained subjects.

(4) From the data obtained, relative strength of elbow joint flexors in the group of the sportsmen correlated negatively with body height and forearm length.

In spite of the small number of experimental subjects we hope that the result may be a stimulus to further evaluation of maximum isometric strength.

References

HETTINGER, T.: Isometrisches Muskeltraining, 2nd ed. (Thieme, Stuttgart 1966).

HILL, A. V.: Dynamic constants of human muscle. Proc. roy. Soc. B. *128:* 263–274 (1940).

IKAI, M. and STEINHAUS, A.: Some factors modifying the expression of human strength. J. appl. Physiol. *16:* 157–163 (1961).

JOACHIMSTHALER, F.: The influencing of physical efficiency of canoesitis by means of some indices of somatometry and muscular strength. Teor. Praxe těl. Vych., Prague *16:* 518–525 (1968).

NELSON, R. C. and FAHRNEY, R. A.: Relationship between strength and speed of elbow flexion. Res. Quart. amer. Ass. Hlth phys. Educ. *36:* 455–463 (1965).

NEUMAN, J., SUKOP, J., and JOACHIMSTHALER, F.: Objective evaluation of starting movements and determination of factors influencing their execution, in v. Sborník VR ÚV ČSTV, 4, pp. 133–151 (Olympia, Prague 1968).

TORNVALL, G.: Assessment of physical capabilities. Acta physiol. scand. *58:* 1–55, suppl. 201 (1963).

WARTENWEILER, J.: Eine Testbatterie zur Erfassung von Grundeigenschaften der menschlichen Motorik. Biomechanics I, 1st Int. Seminar Zurich, 1967, pp. 273–277 (Karger, Basel 1968).

Authors' address: Dr. JAN NEUMAN and Dr. IVAN KUTA, Physical Education Research Institute, *Prague* (Czechoslovakia)

Medicine and Sport, vol. 6: Biomechanics II, pp. 129–132
(Karger, Basel 1971)

The Influence of the Tempo of Playing on the Rhythmic Structure Studied at Pianist's Playing Scales[1]

CH. WAGNER

Max-Planck-Institut für Arbeitsphysiologie (Director: Prof. D. W. LÜBBERS), Dortmund

Among the technical abilities necessary for piano playing there is one of great importance: that is to *play quickly and with rhythmical precision.* That does not mean that I want to reduce the pianist's art to this ability. Considering, however, the literature of etudes of the last 200 years, we notice that almost all the exercise material is devoted to this aim: to reproduce at a quick tempo and in a precise rhythm scales, thirds, arpeggios etc., which are noted in equal note values.

Practice knows quite a number of different methods of acquiring this ability. The most widespread method is that of beginning a piece at an extremely slow tempo, about 2 or fewer strokes per second and to accelerate the tempo by very small degrees through weeks or months. Occasionally, however, the opposite opinion is sustained: Playing quickly is said to be something fundamentally different from playing slowly and, therefore, hardly to be acquired by slow previous exercises. In view of this difference of opinions we studied the problem of how the tempo of playing influences its rhythmic structure with someone who is master of the instrument – the professional pianist.

Method

The characteristic features of the piano cause the rhythmical structure of the playing — disregarding the rests in the musical text — to depend on the time intervals between the beginnings of successive notes. Using electric impulses, we can record these intervals by marking the moment of the hammer touching the string, recording the sequences of impulses and measuring the time intervals between the impulses. The

1 This study was supported by the Stiftung Volkswagenwerk.

measurement error of our experimental arrangement including the Bechstein piano used is in the ±1 msec range.

The test consisted in playing with the right hand a C major scale over two octaves by using the usual finger set: 1 2 3 1 2 3 4. The 8 tempo graduations were 2.0, 3.1, 4.0, 6.1, 8.0, 10.1, 12.3, and 13.9 strokes per second, each marked by a metronome in advance of each new exercise. At the end of the 8th exercise the subject was allowed to choose an even faster tempo, without this being marked beforehand.

The group of 11 subjects – the series is being continued – consisted of 5 male and 6 female pianists aged between 19 and 55, partly pianists who have practically proved their technical ability in concert and radio performances and by gramophone records, partly students of a Music Academy at the stage of concert maturity.

Results

As a measure of the rhythmical precision we chose the deviation of the single intervals from the average interval duration. The first and final intervals were excluded because they were often prolonged. The results can be summed up as shown in figure 1. In spite of the differences of subjects as regards age, sex and pianistic experience, to which may be added different anatomical-physiological conditions, their performances of this model task show equal responses on essential points: the optimum low scatter (±6–9 msec) in the medium tempo range and almost symmetrically increasing deviations in the direction of slow as well as of quick playing. The difference in limit frequencies attained by the different players is noticeable. It appears, however, that highly trained pianists playing scales generally attain values of 13–14/sec, although with an individually very different deviation range of the single intervals. With this frequency of strokes the tempo specifications of musical literature may be realized in most cases.

Discussion

Practically more important than the investigation of the *limit of the frequency of strokes* is the one studying the conditions under which an *increase in the frequency of strokes* is attained up to that stage which is required even in music not expressly meant for virtuosos. Already the high deviation rates at slow playing (fig. 1) make it doubtful whether slow previous exercises promote the attainment of a high playing speed. Only when we analyze the single values *in their temporal sequence* can we see the fundamental difference between periodical finger movements at a slow

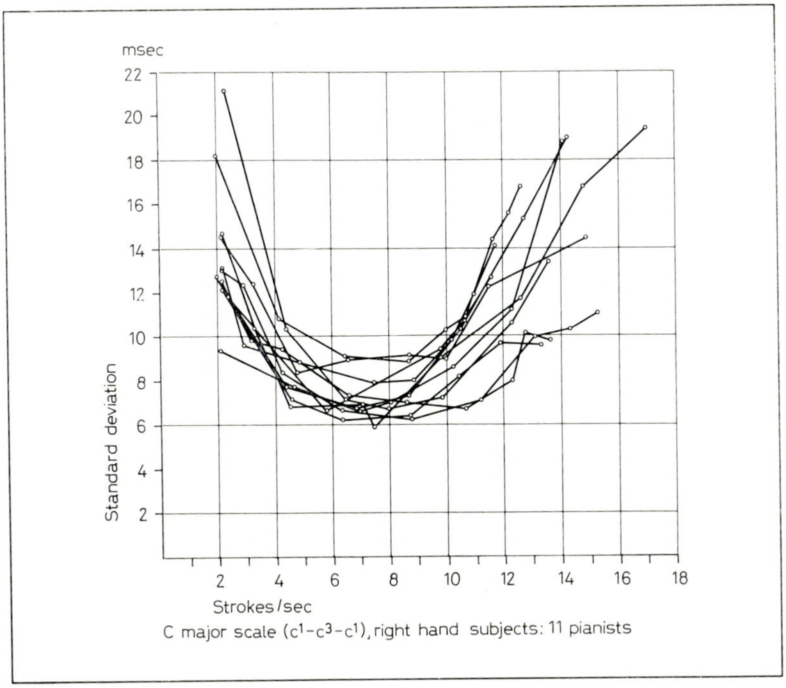

Fig. 1. Playing of C major scales on the piano; rhythmical precision as a function of the tempo of playing.

and those at a fast speed. The performance of quick passages cannot be realized without extensive automation of at least partial processes of movement. Former investigations [WAGNER, 1968] had shown that the repetition of tone sequences leads to time-interval patterns. These have to be regarded as an expression of this very automation and their degree depends on the level of training. It is evident that *the tempo of playing has an important influence on the forming of such automatism.* In figure 2 one of the 11 cases demonstrates what as a rule was found in all: Whereas at a slow playing (fig. 2, top) the 3 passages do not resemble each other very much, they show a relatively high degree of conformity in the two other passages (fig. 2, middle and bottom). This result is the more astonishing as the sequence of the time intervals, different from the preceding test tasks, is considerably longer and the changing of fingers is much more complicated. One gains the impression that with the first performance a motor program is stored which *as a whole* remains dis-

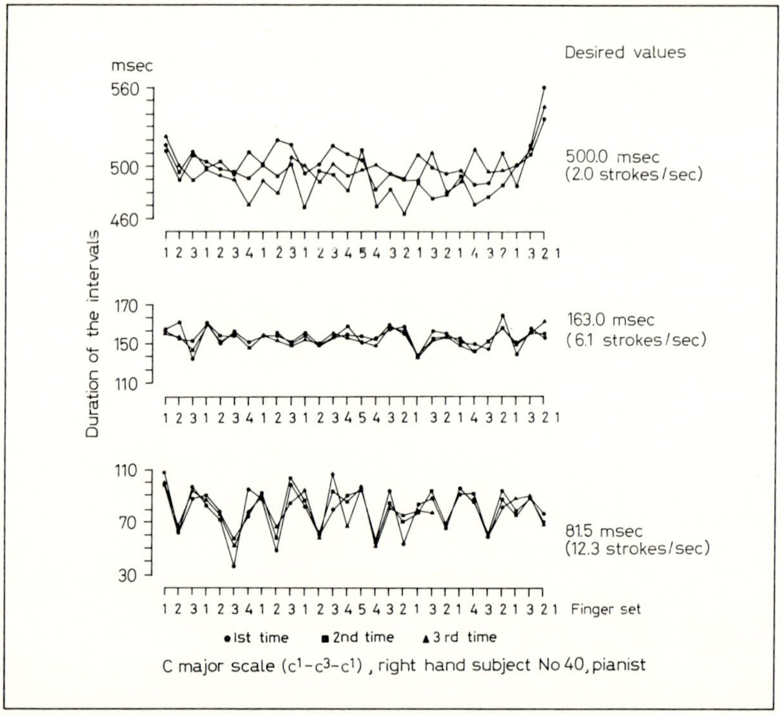

Fig. 2. Playing of C major scales on the piano at a slow, middle and quick tempo; time intervals of the successive notes, two repetitions each time.

posable for some time. It seems, however, that for the formation of such a program *a certain minimum performance speed* is required. With regard to the problem of motor learning it would be of great importance to know what extent such movement programs can have, how exactly and for what duration they are fixed, if there is a certain critical speed which promotes the automation of larger movement units and which is valid not only for a special motor task but generally, and finally what individual differences exist in this field. We will examine these questions in further investigations.

Reference

WAGNER, CH.: Untersuchungen zur Ergonomie des Klavierspiels. Biomechanics. I. 1st int. Seminar, Zürich 1967, pp. 246–270 (Karger, Basel 1968).

Author's address: Dr. med. CHRISTOPH WAGNER, Max-Planck-Institut für Arbeitsphysiologie, Rheinlanddamm 201, *D-46 Dortmund* (Germany)

IV. Kinetics of Human Motion

a) Muscular Properties

Medicine and Sport, vol. 6: Biomechanics II, pp. 134–145
(Karger, Basel 1971)

Basic Kinetic Rules for Simple Human Movements

J. Wartenweiler and A. Wettstein

Department of Physical Education of the Swiss Federal Institute of Technology, Zürich

Introduction

In this investigation a series of simple movements are graphically presented as a means of formulating basic kinetic rules of movement. These rules with certain modifications should also apply to more complex movements.

Horizontal Swinging of the Arms as a Test Movement

In the present study we especially analysed the horizontal swinging of the arms (fig. 1). It is a simple movement with two antagonistic groups of abductors and adductors. The arm of the test person was 72 cm long from the rotation centre in the shoulder joint to the end of the closed hand; 43 cm were calculated as inertia radius. The weight was 4.5 kg.

The left arm was also swung as a counter movement to the right arm. Thus any influence of the movement by a turning of the trunk was excluded. The deflection of the angle of the arms with regard to the shoulder line varied during measurements between 150° and 30°, i.e. positions between an approximately straight line and the crossing of the arms. Tested was above all the angle position of 90° when the arms were stretched forward horizontally.

The static force was determined by an electronic pressure and pull measuring device and measured at the level of the inertia radius. The measured values are re-

The angle was measured with a potentiometer which was attached to a light metal frame on the right shoulder joint.

To determine the tangential acceleration at the mass center of the arm a Statham accelerometer A 6-25 was attached at the radius of inertia.

The dynamic force was calculated by multiplying the mass of the limb by the previously described acceleration. This force can be directly compared with the measured static force.

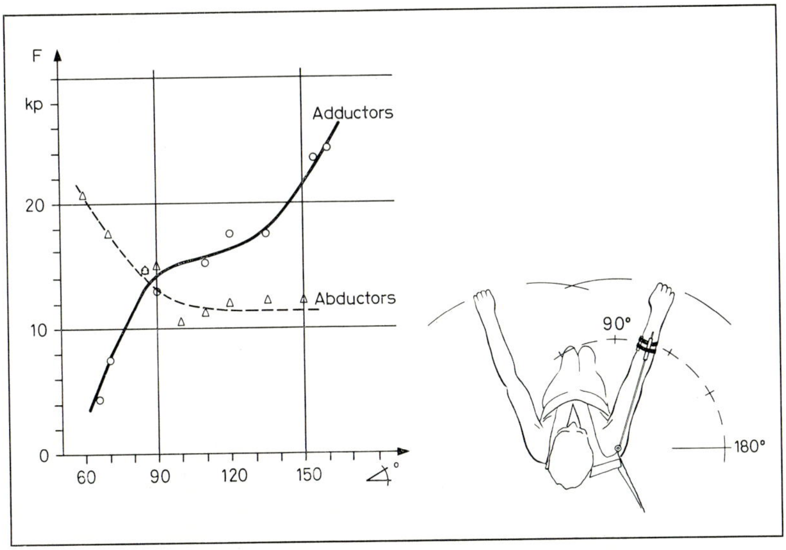

Fig. 1. Isometric maximum force of the abductors and adductors of the shoulder joint in relation to the position of the arm. Measurement at the radius of inertia.

The acceleration as well as the static increase of the force, respectively the static force level (IMF = isometric maximum force), was recorded as shown in figures 3a-c, 5, 7 and 9, enabling a visual comparison of the static and dynamic forces to be made.

Real Muscular Force and Mechanically Operative Force

In the evaluation of the force in action the following general principles must be taken into consideration:

(1) The mechanical effective force is dependent on the lever arm of the acting force and on the geometry of the joint.

(2) The static muscular force is dependent on the length of the muscle. More force is produced in the lengthened state than when the muscle is shortened.

(3) In isometric muscle activity the maximum force is achieved after approximately 400 msec at the earliest (fig. 2).

Explanation: The time that elapses till the full development of the maximum force is partly dependent on the process of innervation and concentration, and partly on the elasticity of the tissue.

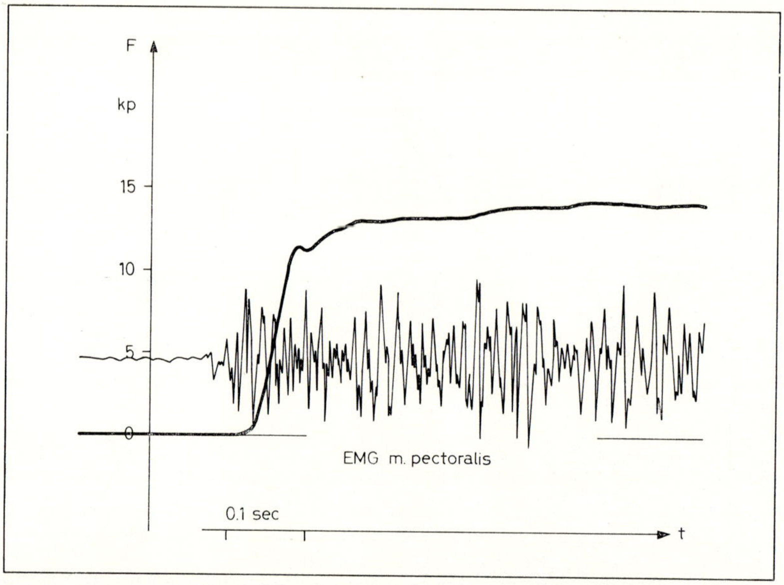

Fig. 2. Quickest rising to the isometric maximum force (IMF).

(4) The dynamic force is dependent upon the contraction process. As the speed of the contraction increases the mechanically effective force decreases in concentric and increases in eccentric muscle activity. This is due to the time required for the contraction process, to the chemomotoric conditions and to the attenuating quality of the muscles ('inner friction', 'viscosity'), as well as to reflex influences.

Basic Kinetic Rules

1. Aperiodic Movements in One Direction

Rule 1. Two forces are required for swinging movements in one direction: the accelerating force and the decelerating force. These forces are displayed during the whole process with regard to both space and time (fig. 3a, b, c).

Explanation: The more evenly the force is distributed the lower is the force peak. High force peaks result in high tissue demands and may lead to injuries.

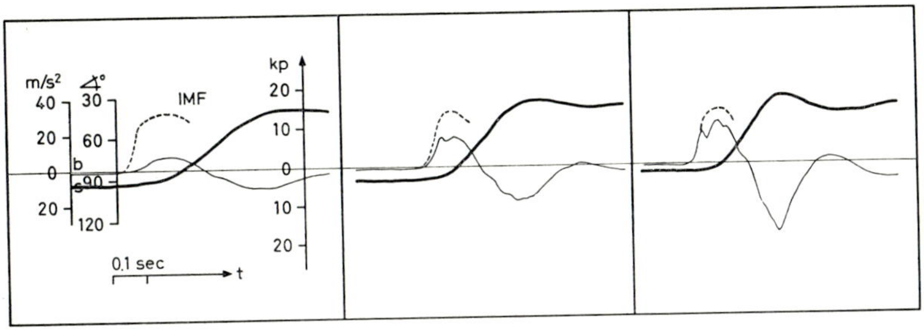

a *b* *c*

Fig. 3. Distribution of force when swinging the arm horizontally inwards. *a* Slow, *b* fast, *c* maximum speed. b = Acceleration in the radius of inertia, s = shoulder-joint angle, IMF = isometric maximum force. Direct comparison of acceleration (m/s^2) and force (Kp) is possible because of the constant arm mass of 4.5 Rg.

Rule 2. Accelerating and decelerating forces develop by continuous increasing and decreasing (fig. 3a, b, c).

Explanation: To keep the peaks of force as low as possible, the distribution of the force should have a rectangular form. Not only high force peaks but also sudden force increases cause a great stress on the tissue. For this reason the optimum between the gradual increase and the lowest possible force peaks is being aimed at.

Rule 3. The higher the input of force, the higher the height of the force peak and the steepness of the force increase (fig. 3a, b, c).

Rule 4. With concentric muscle activity (movement with muscle shortening) the static maximum force cannot be achieved (figs. 3c and 4).

Explanation: (1) The movement of the limb, in the same direction of muscle contraction, causes a delay in the development of the force. (2) The mechanical effect of the muscle force reduces by the amount of the 'inner friction' which has to be overcome in the contraction. (3) In fast short movements the path is run in so short a time that maximum force cannot be achieved.

Rule 5. With eccentric muscle activity (movement with muscular lenghtening) supramaximum forces (i.e. forces which exceed the static maximum force) can be achieved (figs. 4 and 5).

Explanation: (1) The movement of the limb, in opposite direction to muscle tension, causes precipitation of the development of the force. (2) 'Inner frictional forces' are added to the muscular force.

Note: In eccentric as in concentric muscle activity the path may be run before the static muscle force is achieved.

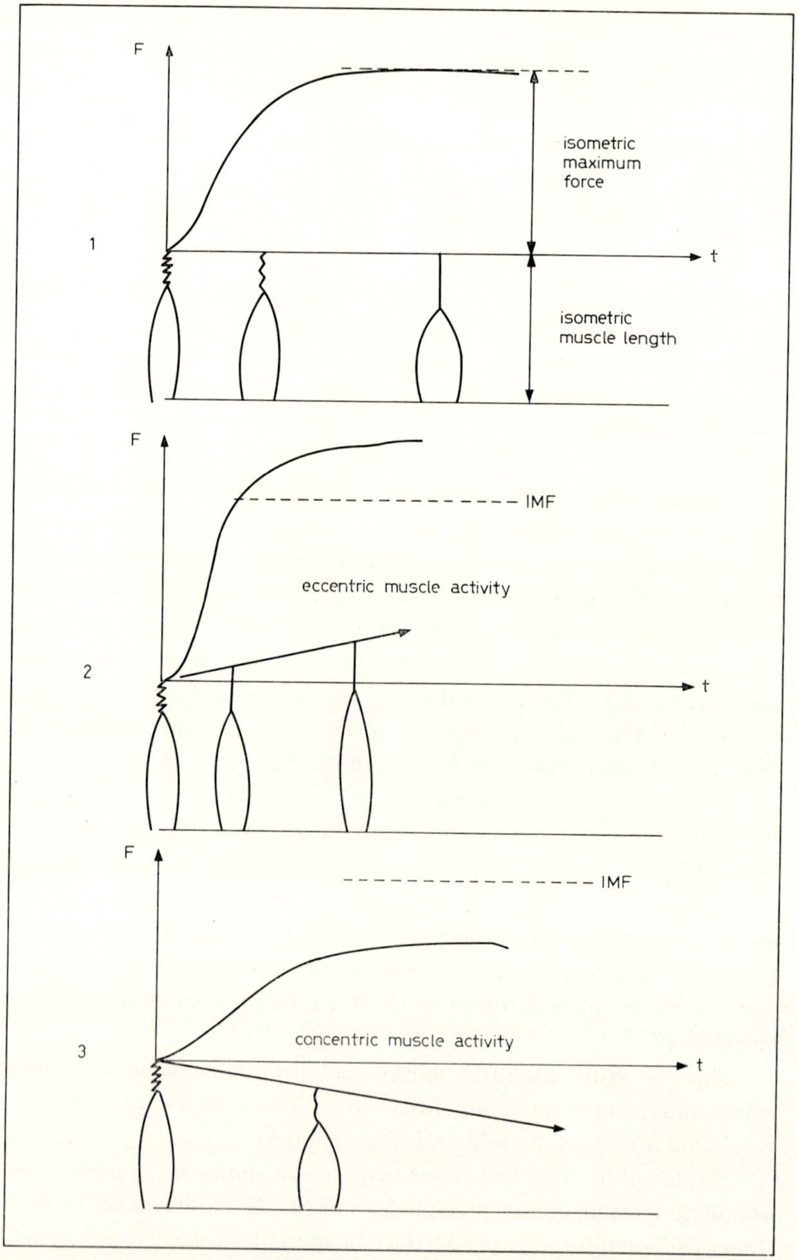

Fig. 4. Scheme for the development of maximum muscular force. *1* Isometric, *2* eccentric, *3* concentric.

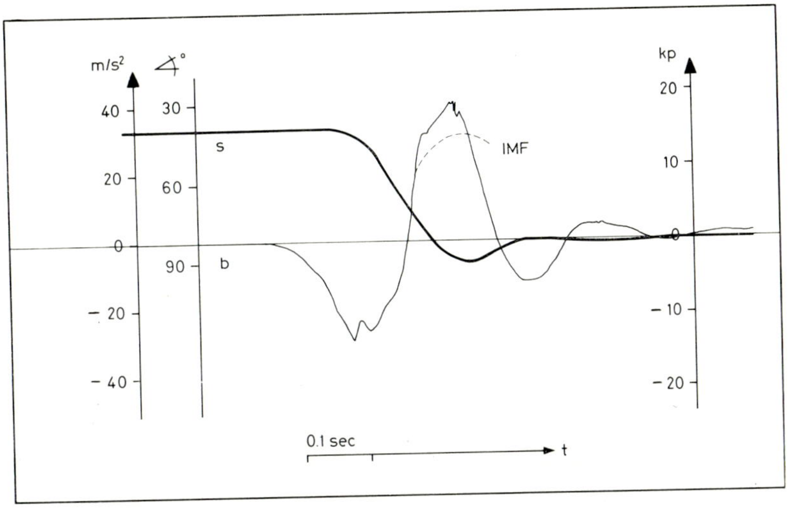

Fig. 5. Distribution of force when accelerating and decelerating the arms horizontally outward to a full stop. Notice the difference between decelerating force and isometric maximum force (IMF).

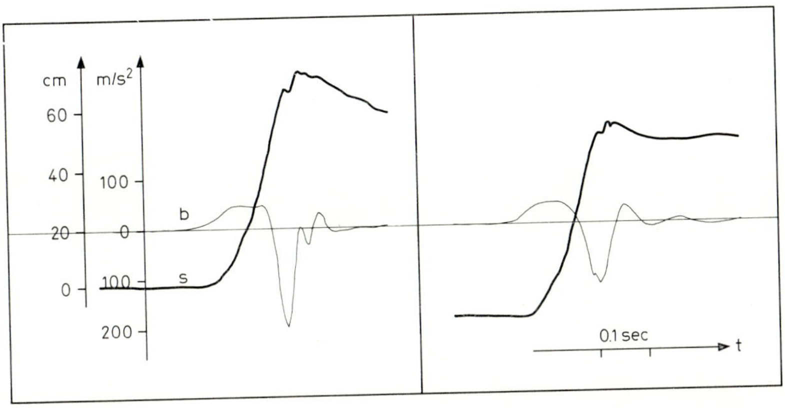

a, b

Fig. 6. Horizontal boxing thrust. *a* Decelerated in full extension of the arm, *b* decelerated before the full extension of the arm. b=Acceleration hand, s=path of the hand.

Rule 6. In final positions of movements, for instance stretching of the arms, high reaction forces can develop which are added to the muscular forces (fig. 6).

Rule 7. Swinging movements with a greater moment of inertia can result in a force plateau near the maximum force.

Explanation: With greater moment of inertia the movement is relatively slow so that it approaches static conditions.

2. Aperiodic Movements Back and Forth

Rule 8. In aperiodic movements back and forth 4 forces are to be distinguished and the middle forces normally form a unit (fig. 7 and 8):

(1) the accelerating force of the backswing,
(2) the decelerating force of the backswing,
(3) the accelerating force of the main movement,
(4) the decelerating force of the main movement.

If a pause occurs between the backswing and the main movement, the middle forces appears separately (fig. 8a).

Rule 9. Movements with a backswing combine the characteristics of eccentric, isometric and concentric muscle activity; more exactly the supramaximum forces become the stronger before and at the point of inflexion

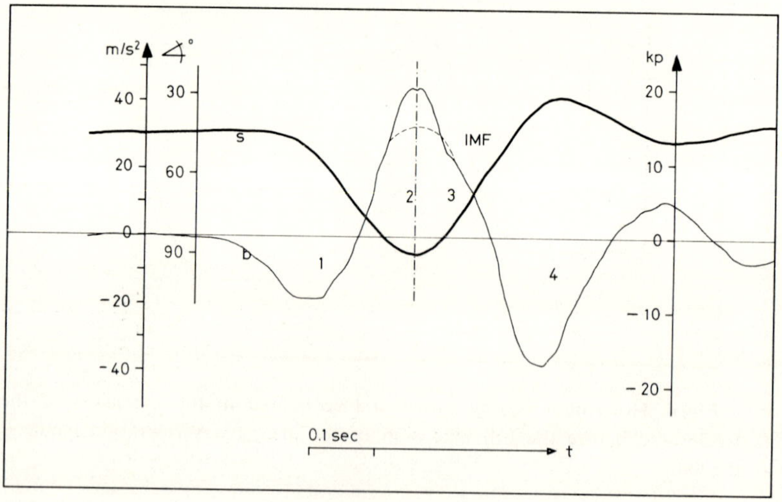

Fig. 7. Fastest aperiodic swinging of the arm outwards-inwards. *1* Acceleration outwards from initial position, *2* deceleration of the swinging outwards, *3* acceleration inwards, *4* deceleration to a stop.

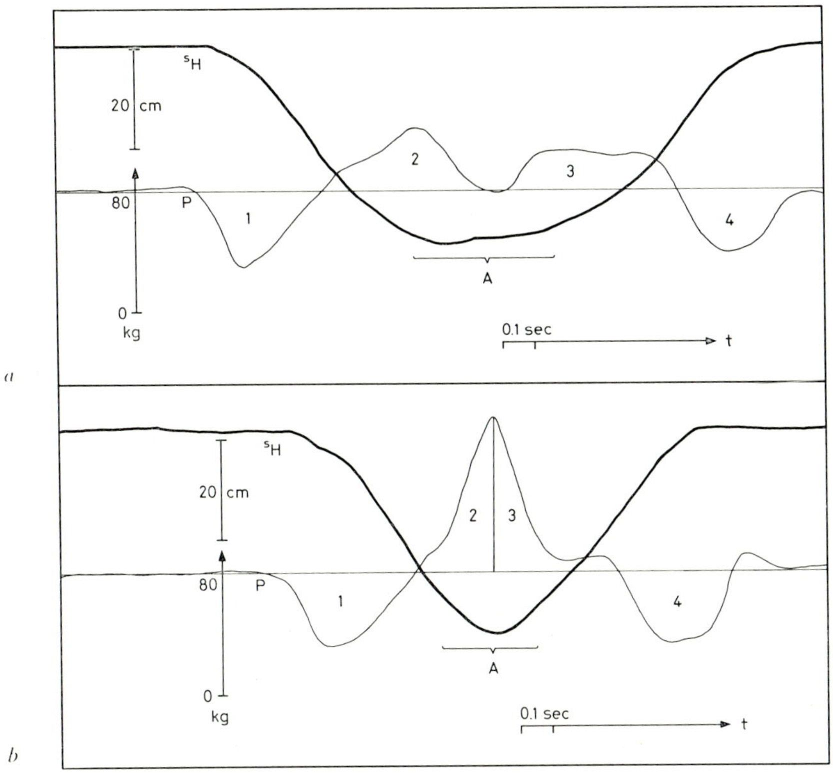

Fig. 8 a. Deep knee flexion. Inflexion with a momentary interruption. A = Voluntary delay in the inflexion.

b. Deep knee flexion. Elastic inflexion of the movement. *1* Decrease of pressure when bending, *2* increase of pressure when braking the falling body, *3* decrease of the positive pressure when stretching, *4* decrease of pressure when in the second phase of the stretching. A = Well-performed inflexion.

the greater the velocity of the movement to be decelerated. In reverse, the force diminishes the faster after the point of inflexion the greater the velocity of the movement becomes (fig. 7).

Rule 10. With a backswing the velocity of the movement can be increased.

Explanation: From the decelerating force of the backswing results an initial force at the point of inflexion which benefits the main movement. (The higher the operative accelerative force in a given course the greater

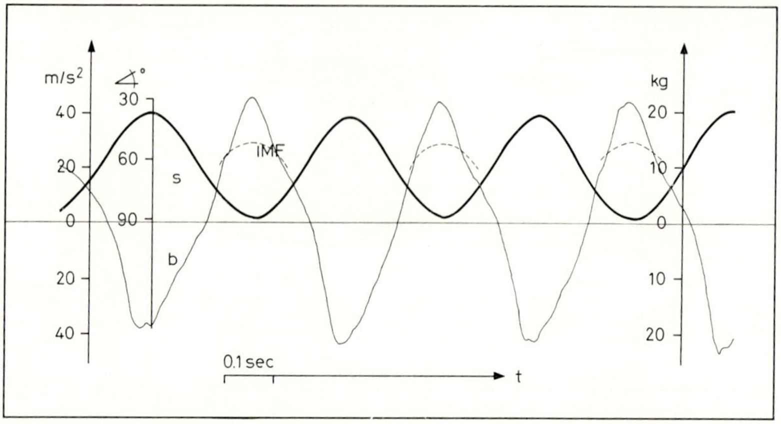

Fig. 9. Periodic swinging of the arms horizontally with maximum force.

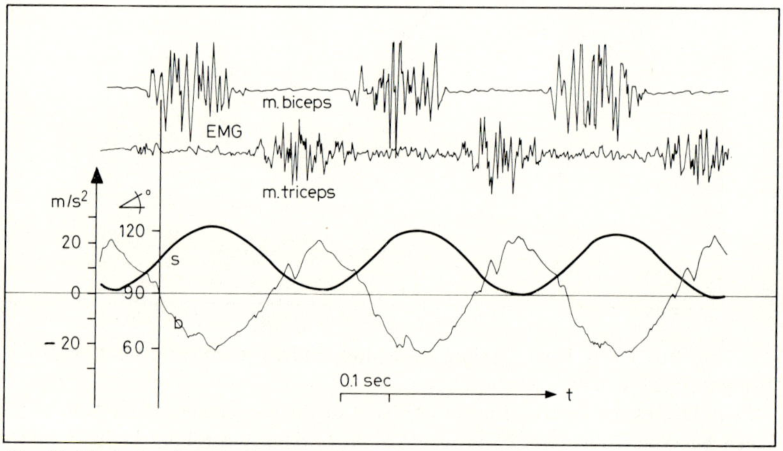

Fig. 10. Horizontal periodical swinging of the forearm.
b = Acceleration of the middle of the hand, s = angle of the elbow.

Fig. 11. Well- and ill-regulated forms of arm swinging.

a EMG: released movement with alternating input of force of the antagonists.

bHand: regular acceleration curve of the hand (the thinly recorded curve in the middle of the picture = tangential acceleration of wrist joint) corresponds to a well-regulated input of forces.

sHand: the path of the hand (the accentuated curve in the middle of the picture = shoulder-joint angle) has a regular form.

b EMG: tense movement in which the antagonists interfere with each other.

bHand: 'Wild' acceleration outbursts show an ill-regulated input of forces with lack of output of forces.

sHand: During the path of the hand the disturbances are less evident than during acceleration.

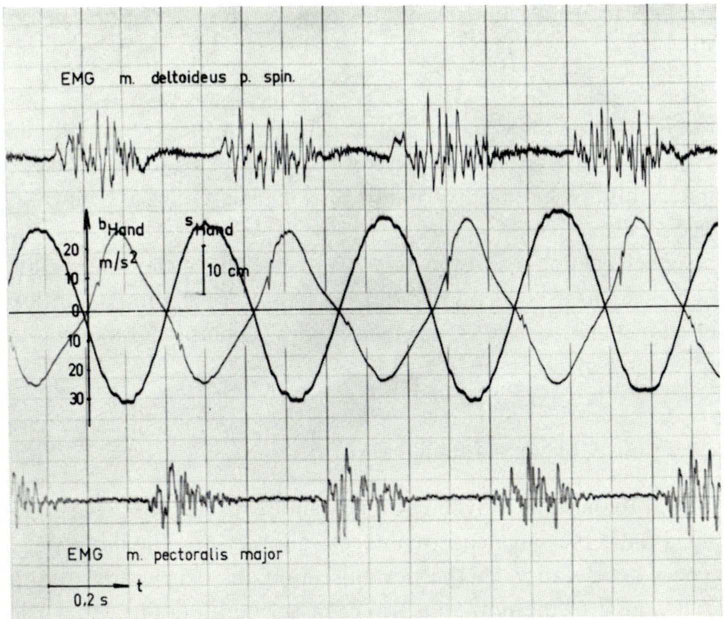
EMG m. deltoideus p. spin.
bHand
sHand
m/s2
10 cm
20
10
0
10
20
30
EMG m. pectoralis major
t
0,2 s

a

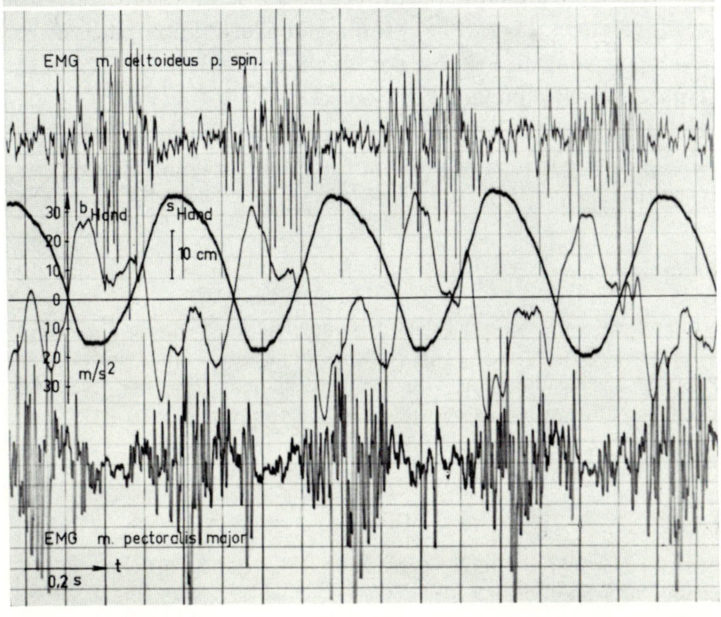
EMG m. deltoideus p. spin.
bHand
sHand
10 cm
30
20
10
0
10
20
30
m/s2
EMG m. pectoralis major
t
0,2 s

b

is the velocity.) The backswing is especially useful in fast short movements which would never reach a high muscle tension by pure concentric muscle activity.

HOCHMUTH [1968] finds for the standing jump an optimum proportion of

$$\varkappa = \frac{\text{deceleration thrust}}{\text{acceleration thrust}} = 0.3 \text{ to } 0.4.$$

Rule 11. An increase in the initial velocity of the movement can also result from a static pretension (flick of the finger); contrarily, movements immediately after dynamic overload are not faster [NELSON and NOFSINGER, 1965].

Rule 12. Reactional forces which are the result of the final position of movements can–as in a deep knee flexion–help to overcome the inflexional point of movement, which is associated with an unfavourable joint position (fig. 8a and b).

3. Periodic Movements Back and Forth

Rule 13. In periodic movements back and forth, the force performs two functions. Firstly the force causes a deceleration during the movement, and secondly an acceleration in the new direction. In decelerating the 'inner friction' is added to the muscular force; in accelerating the 'inner friction' has a negative effect. 'Tooth-like' force and acceleration patterns respectively develop as a result. In these patterns force peaks caused by reactional forces (rule 6) are also included (fig. 9).

Rule 14. In periodic movements back and forth antagonistic muscles contract alternately with a minimum of overlap (in an ideal case with none at all) (fig. 10).

Summary

In the present work we are trying to summarise the principal characteristics of simple movements. Additional research will be necessary and several points still require further clarification. But with greater deviations from the described standards we now have the possibility of evaluating faulty movements (fig. 11).

References

ASMUSSEN, E.; HANSEN, O., and LAMMERT, O.: The relation between isometric and dynamic strength in man. Bull. Féd. int. Educ. phys. *1:* 9–19 (1967).

Basmajian, J. V.: Muscles alive (Williams & Wilkins, Baltimore 1967).

Donskoi, D. D.: Biomechanik der Körperübungen (Sportverlag, Berlin 1961).

Hill, A. V.: The abrupt transition from rest to activity in muscle. Proc. roy. Soc. B. *136:* 399–420 (1949).

Hochmuth, G.: Biomechanische Prinzipien. Biomechanics. I. 1st int. Seminar, Zürich 1967, pp. 155–160 (Karger, Basel 1968).

Komarek, L.: Biomechanics of training of simple rhythmic efforts. Biomechanics. I. 1st int. Seminar, Zürich 1967, pp. 209–213 (Karger, Basel 1968).

Margaria, R.; Cavagna, G. A. e Saibene, F. P.: Possibilità di sfruttamento dell'elasticità del muscolo contratto durante l'esercizio muscolare. Boll. Soc. ital. Biol. sper. *39:* 1815–1816 (1963).

Nelson, R. C. and Nofsinger, M. R.: Effect of overload on speed of elbow flexion and the associated after-effects. Res. Quart. *36:* 174–182 (1965).

Reichel, H.: Muskelphysiologie (Springer, Berlin 1960).

Rodahl, K. and Horvath, S. M.: Muscle as a tissue (McGraw-Hill, New York 1962).

Vredenbregt, J. and Koster, W. G.: Some aspects of muscle mechanics *in vivo.* IPO annu. Progr. Rep. *1:* 94–100 (1966).

Wartenweiler, J.; Steinemann, M. und Wettstein, A.: Über die Kraftentfaltung bei Kontraktionen der Skelettmuskulatur. Verh. Schweiz. naturforsch. Ges. *142:* 138–140 (1962).

Wartenweiler, J. und Wettstein, A.: Untersuchung und Analyse der Sägebewegung. Int. Z. angew. Physiol. *21:* 69–73 (1965).

Wartenweiler, J. und Wettstein, A.: Die Sägebewegung bei variabler Sägegeschwindigkeit. Int. Z. angew. Physiol. *22:* 181–184 (1966).

Wartenweiler, J.; Wettstein, A., and Lehmann, G.: Shot put, subjective and objective aspects of motor design. Proc. 2nd int. Congr. Sportpsychol., Washington 1968.

Wartenweiler, J. und Wettstein, A.: Biomechanische Grundprinzipien für schwunghafte Bewegungen. Körpererziehung *7:* 149–160 (1965).

Wilkie, R. D.: The relation between force and velocity in human muscle. J. Physiol., Lond. *110:* 249–280 (1949).

Author's address: Prof. Dr. J. Wartenweiler, Abteilung X für Naturwissenschaften, Eidgenössische Technische Hochschule, *CH-8000 Zurich* (Switzerland)

Medicine and Sport, vol. 6: Biomechanics II, pp. 146–153
(Karger, Basel 1971)

Elasticity of Skeletal Muscle in Relation to the 'Sliding Filaments' Model

T. Blangé, J. M. Karemaker, and A. E. J. L. Kramer

Laboratory of Medical Physics and Physiological Laboratory, University of Amsterdam, Amsterdam

In relation to studies on human movements it was felt to be of interest to extend existing knowledge of the mechanical behaviour of active skeletal muscle.

Starting from the known structural properties of the muscle it is possible to construct a model which to a certain extent explains the mechanical events. A successful model based on this procedure is that of the sliding filaments, put forward by A. F. Huxley and Niedergerke and by Hanson and H. E. Huxley [as quoted by A. F. Huxley, 1957].

To study the validity of this theory for whole muscle preparations, measurements were made of the elasticity of rat soleus muscle. To facilitate a better understanding of the experimental approach chosen, a short introduction is necessary.

It is generally accepted that there are two kinds of filaments in skeletal muscle, one consisting of myosin and the other of actin, arranged as shown in figure 1. In active muscle, projections from the myosin filament can form links between the myosin and the actin filament. It is assumed that muscle force is generated within these projections after the link has been made,

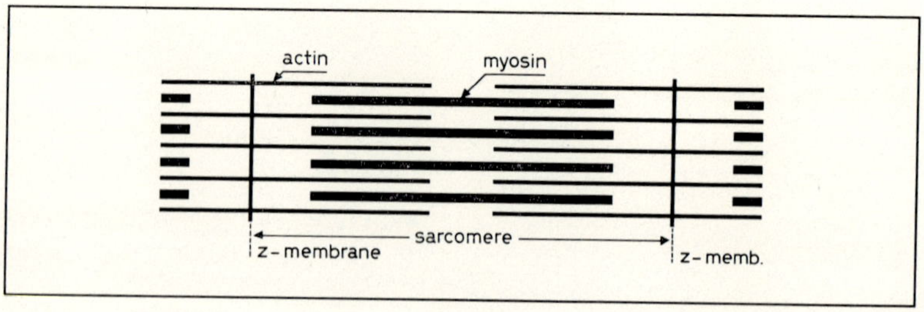

Fig. 1. The arrangement of filaments is shown in a special longitudinal section. In reality, each actin is surrounded by three myosin filaments and each myosin by six actin filaments.

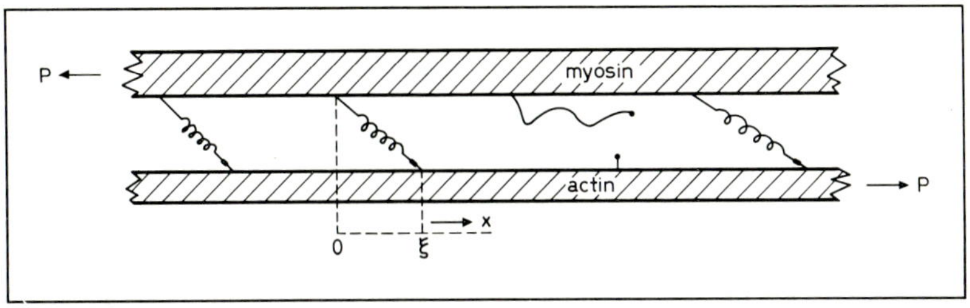

Fig. 2. Scheme of the force generating system. The z-membrane from which the actin filament arises has to be located on the right side of the figure. The connection-points of the projections are taken at regular distances along the myosin as well as along the actin filament [HUXLEY, 1957].

for instance by their transition into a coiled elastic configuration [DAVIES, 1963]. The isometric length-tension relationship that is found [GORDON, HUXLEY and JULIAN, 1966] can easily be explained if one assumes that the tension is proportional within a certain range to the length of the overlap zone of the myosin and actin filaments. The number of projections that can actually form the links should be proportional with that same length.

The model was further adapted by HUXLEY [1957] by means of the following simplifying assumptions.

(1) In a first approximation it was assumed that the filaments are rigid. The force generating system can then be represented as shown in figure 2. The force exerted by one link in the x-direction is $P_l(\xi)$, where ξ is the distance in the x-direction between the points where a link is connected with the filaments. The total muscle force will in this way be determined by the dependence of P_l on ξ, and by the distribution of the links as a function of ξ. This distribution will be represented by $n(\xi)$, the number of links, or rather the probability of finding a link in a half-sarcomere with $x < \xi < x + dx$. (The total number of links — not the number of projections — in the half-sarcomere is then $\int n(\xi)d\xi$.)

(2) $P_l(\xi) = S_l \cdot \xi$. This implies that the link acts as an ideal spring.

(3) In the case of an isometric stationary force, $n(\xi)$ has the shape indicated in figure 3A. When a half-sarcomere shortens instantaneously, say by an amount $\triangle x$, the resultant $n(\xi)$ is assumed to have the shape shown in figure 3B or possibly 3C, if for some reason the links cannot pull in the 'wrong' direction. The relationships between the force P and the shortening and between the stiffness $S = dP/dl$ and P for the sarcomere (and for the whole muscle considered as a number of sarcomeres in series) will then be

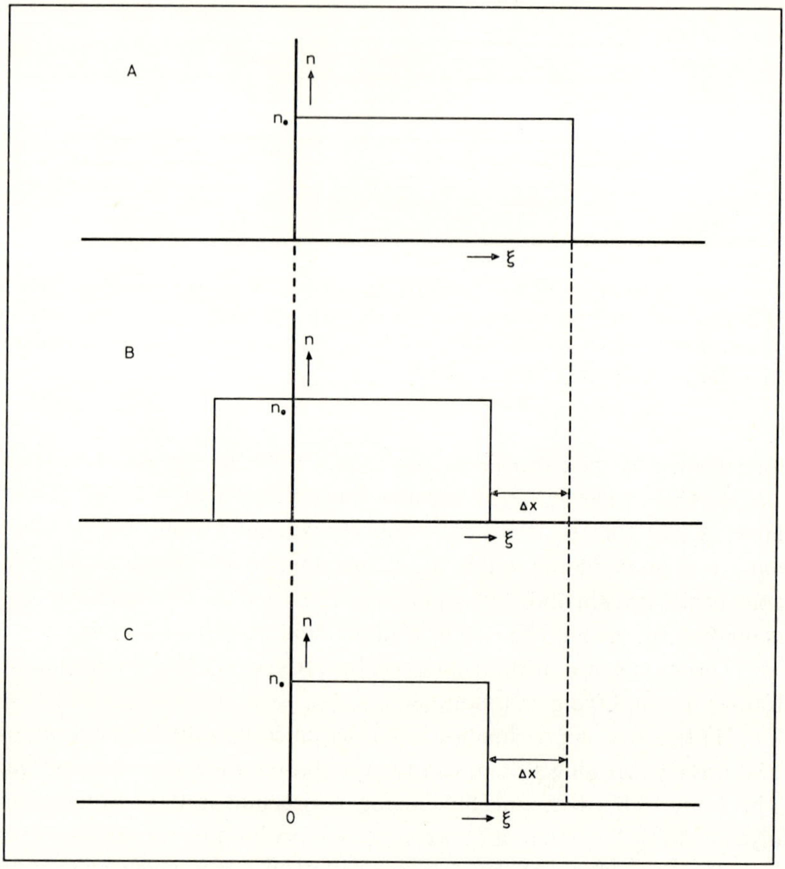

Fig. 3. For isometric conditions a simple distribution $n(\xi)$ as proposed in the text is given in A. Shortening of the half-sarcomere by an amount $\triangle x$ decreases the value of ξ of a link to $\xi - \triangle x$, resulting in the distribution B or C (see text).

of the shapes shown in figures 4A and 4B, which correspond with figures 3B and 3C respectively.

One can also try to vary P, not by shifting the distribution, but by altering its height, i.e. the value of n_0 in figure 3. This can be done by submaximal stimulation or by applying lower stimulation frequencies. Stiffness can then be calculated again as a function of P and since both stiffness and force are proportional to the number of links, or to n_0, it is clear that in this case S is proportional to P (fig. 4C).

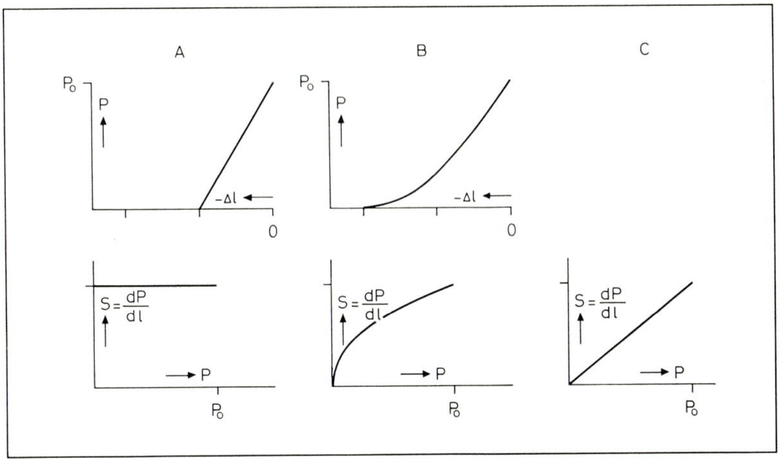

Fig. 4. The distribution of figure 3B leads to a force-shortening and a stiffness-force curve as in A. In the same way 3C leads to B in which case the force-shortening curve is a parabola and stiffness is proportional to the root of P. The stiffness-force curve resulting from adjustment of the force by means of stimulation is shown in C.

As shown in the above-mentioned example, a general consequence inherent to this model is that the stiffness of a muscle depends on the way its force is brought about.

The experimental set-up used to determine what difference exists between the two stiffness-force curves was as follows. The soleus muscle of the rat was chosen for the experiment because it seemed well adapted to our problems and because extensive data about the rat soleus muscle have been published by CLOSE [1964]. The proximal side of the muscle with the blood and nerve supply was left intact inside the animal and the tibia was clamped to the apparatus. The distal tendon was tied directly to a force transducer. The muscle was stimulated via the tibial nerve. As in all measurements mentioned here, muscle length was chosen for maximal isometric twitch performance, for which fibre length was about 12 mm. The stiffness in the state of rest was then negligible.

At a preset time after the beginning of stimulation, the force transducer was given a sudden displacement, which allowed the muscle to shorten. This displacement was measured with another transducer. Figure 5 shows schematically the mechanical part of the apparatus, but we will not go into further detail here.

Figures 6 and 7 show typical examples of the outputs of the two transducers during a measurement with a different time scale. Accurate measurement of the force was possible 1.5 to 2.5 msec after the beginning of

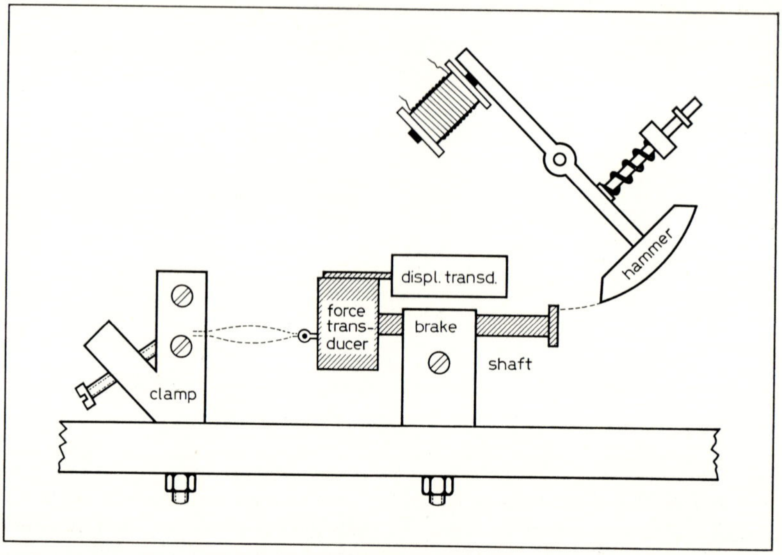

Fig. 5. Schematic representation of the apparatus. At a preset time after the beginning of stimulation, the electromagnet is switched off and the hammer is shot away by the spring. Some milliseconds later the hammer taps against the shaft and the hatched part including the force transducer takes over the impulse of the hammer. The velocity of the force transducer is then decelerated to zero within 1 msec by means of the (dry friction) brake.

the displacement which itself was faster but induced some mechanical resonances of the system.

The force-shortening curve measured with this device is given in figure 8. P is measured as soon after the shortening as feasible. It should be noted that the error in this curve in the range of the large displacements (low forces) can be considerable for the following reason. Since the displacement in this range is large, the 'non-accurate' time of the force transducer is relatively long. The force exerted by the muscle after the displacement may become very low or even zero, but then the muscle may rebuild a great deal of force. However, the (isotonic) shortening velocity of the muscle will not exceed 100 μm/msec, limiting the inaccuracy as a result of the 'non-accurate' time. The maximum correction that can be expected is of the order of 250 μm. The error is much less in the range of the high forces, thanks to the hyperbolic nature of the force-velocity curve [CLOSE, 1964].

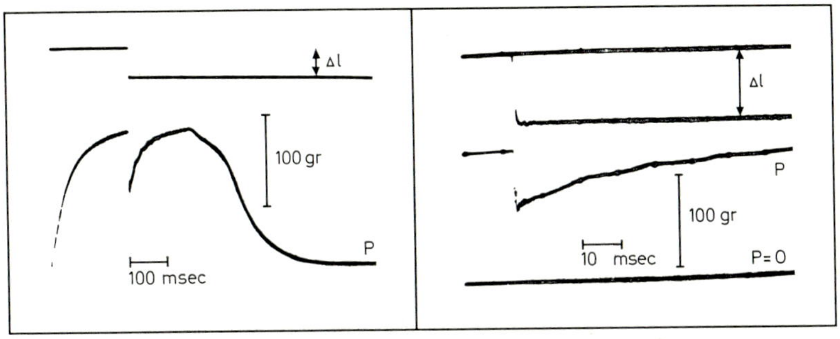

6, 7

Fig. 6. The upper trace shows the output of the displacement transducer. The displacement was about 160 μm. The lower trace shows the output of the force transducer. It has some creep, for which the jump in the force has to be corrected. This is done in the results shown in figures 8 and 9.

Fig. 7. The upper and lower traces represent displacement (about 150 μm) and force, respectively, as in figure 6. The quality of the original picture was better, so that the effect of the acceleration and deceleration phase of the displacement on the force transducer could be seen more clearly.

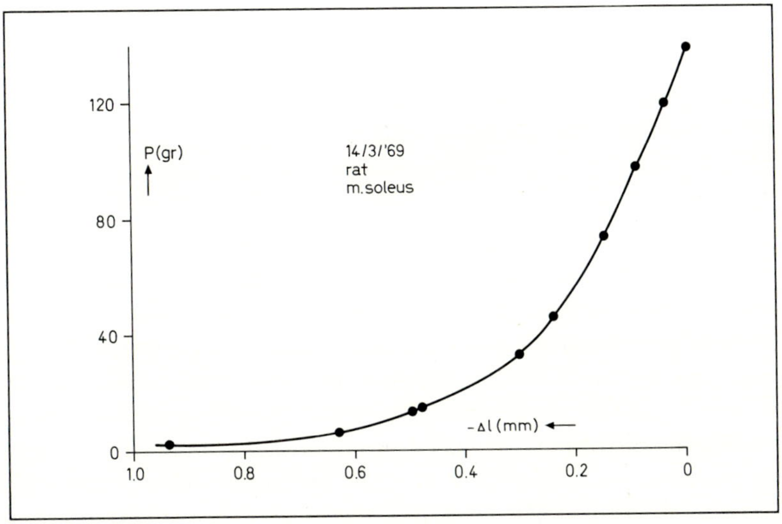

Fig. 8. Force-shortening curve of the soleus muscle of the rat, measured as described in the text.

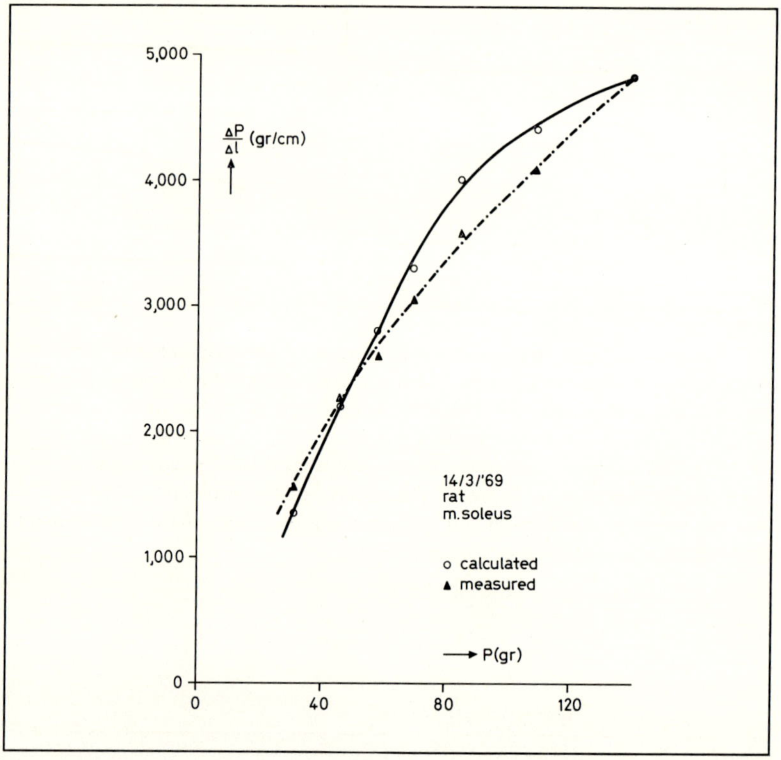

Fig. 9. The broken line shows stiffness, measured by means of a displacement of about 40 μm, as a function of P, the force just before the displacement. The solid curve was derived from figure 8 (see text).

The broken line in figure 9 was measured at different forces obtained by changing the frequency and by using a sudden shortening. In the ideal case the shortening of the muscle should be infinitesimally small. From figure 8 a displacement of 40 μm was considered an appropriate compromise between the inaccuracy inherent to a finite displacement and the inaccuracy of the force transducer.

The points of the 'calculated' curve in figure 9 are derived from figure 8 for the same values of P as in the 'measured' curve, with $S=(P-P')/\triangle l$ and $\triangle l=40$ μm. The value P′ is the force corresponding with a shortening of 40 μm more than that belonging to P. Both curves may thus be directly comparable. They show a marked difference in the range of the high forces. It should be mentioned here that the error discussed with respect to the force-shortening curve tends to give too low values for the stiffness in the 'calculated' curve. A correction, if possible, would, therefore, only increase the difference between the two curves of figure 9.

It might be expected that applying submaximal stimulation instead of lowering the stimulation frequency approaches more the idealised behaviour referred to in figure 4C, because then only part of the muscle is stimulated, probably with the same number of links in the activated fibers as in the case of supramaximal stimulation. The experiment shows a tendency in that direction, but only to a slight extent.

Although some details need further clarification it may be concluded that the results are compatible with the proposed model and assumptions.

However, the results do not irrefutably reject a model of a force source with passive elastic and viscous elements, but only at the cost of a complicated combination of non-linear and linear elements.

In view of the accumulated evidence in favour of the sliding filaments model, it seems justifiable to base further experimental work on this model and to try to adapt it for quantitative analysis.

References

CLOSE, R.: Dynamic properties of fast and slow skeletal muscles of the rat during development. J. Physiol., Lond. *173:* 74–95 (1964). – The relation between intrinsic speed of shortening and duration of the active state of muscle. J. Physiol., Lond. *180:* 542–559 (1964).

DAVIES, R.E.: A molecular theory of muscle contraction: calcium-dependent contractions with hydrogen-bond formation plus ATP-dependent extensions of part of the myosin actin cross-bridges. Nature, Lond. *199:* 1068–1074 (1963).

GORDON, A.M.; HUXLEY, A.F., and JULIAN, F.J.: The variation of isometric tension with sarcomere length in vertebrate muscle fibres. J. Physiol., Lond. *184:* 170–192 (1966).

HUXLEY, A. F.: Muscle structure and theories of contraction. Progr. Biophys. Chem. *7:* 255–318 (1957).

Authors' address: T. BLANGÉ, J.M. KAREMAKER, and A.E.J.L. KRAMER, Laboratory of Medical Physics and Physiological Laboratory, University of Amsterdam, *Amsterdam* (Netherlands)

Medicine and Sport, vol. 6: Biomechanics II, pp. 154–158
(Karger, Basel 1971)

Determination of Muscular Compliance in the Course of Movement

F. Goubel, S. Bouisset, and F. Lestienne

Laboratoire de Physiologie générale, Faculté des Sciences, Lille

Since Hill [1938], it is usual to consider the contracting muscle as a two-component system consisting of 'active' contractile and 'passive' elastic structures in series. The potential energy stored in this elastic element during the initial contraction can play an appreciable physiological role in the performance of the movement during its release [see Margaria, 1968]. It is, therefore, interesting to know the characteristics of muscular elasticity. These have been studied on human muscle under different conditions by Ralston *et al.* [1947], Wilkie [1950], and Cavagna *et al.* [1968]. They entail the knowledge of the variation of tension developed by the muscle (ΔF) in the course of a variation in length (ΔL), the ratio $\Delta L/\Delta F$ being called the compliance. The purpose of this paper is to present a technique for calculating the compliance of a muscle group in the course of movements performed at *variable* velocities.

Technique and Procedure

The subjects are seated and have to perform small amplitude unidirectional flexions of the right forearm in a horizontal plane. An adjustment of the seat allows of putting arm and forearm in the same plane. The experimental set-up is similar to that used in previous studies. The forearm is fixed into a splint. The latter forms a part of a mechanical system which rotates about a vertical axis. The axis of rotation of the elbow, determined by anatomical criteria, coincides with that of the mechanical system. A back support is used to assure the stability of the shoulder. The hand is maintained in a position of semi-pronation and the wrist is immobilized. In order to change the inertia of the mechanical system, weights can be added at a definite distance from the axis of rotation.

The amplitude of the movements is limited to 20° (0.349 rad) counted from a reference position (45° from full extension). The point at which the movement is stopped is determined visually and indicated by photoelectric cells. The movable mechanical system is equipped with an accelerometer measuring in a tangential direction and a goniometer. Thus, tangential acceleration (Θ'') and angular displacement

(Θ) are continuously recorded during the movement. Using the biomechanical constants a and b, which are the distances from the axis of rotation to the upper and lower insertions of the biceps brachii, it is possible to express the length of this muscle in terms of the angle of the forearm (α) in relation to the full extension. Applying the theorem of the kinetic moment and using the constants a, b and I (moment of inertia of the system), it is possible to calculate the resultant tension F in the direction of the biceps brachii as a function of Θ'' and α. The individual constants a, b and I were experimentally measured [Bouisset and Pertuzon, 1968; Pertuzon and Bouisset, 1970].

The tension (F) and the length (L) of the biceps are calculated instantaneously by means of an on-line analog computer. When the programmation of the formula from which F and L may be derived is set up, it is sufficient to set the values of the constants I, a and b on the potentiometers and to feed the appropriate modulus with the variables Θ and Θ'' after suitable amplification. The surface electrical activity (EMG) of biceps and triceps brachii is picked up and integrated (EMGi) in the form of pips (Q_B, Q_T) following a usual method. An electronic system permits to obtain the 'envelope' of muscular activity (EMGe) by the rectification and filtering of muscular potentials. The myoelectrical activity and the different biomechanical variables are simultaneously recorded by means of a moving magnet oscillograph with appropriate frequency range. An X-Y dual trace oscilloscope permits the recording of the electromyogram and the dynamic tension (F) of muscle biceps as functions of its length (L). A tape recorder offers the possibility of storing some data which can be examined by means of an X-Y data recorder or sent to the analog computer for further operations (fig. 1).

Each of three subjects were examined twice. Each examination consisted of series of movements performed without any weight added to the mechanical system (0 kg) and with the addition of a light weight (3 kg). For each experimental situation, that is, for each inertia, the subject first performed the movements at some selfdetermined velocity. Then, according to the instructions of the experimenter, the movements were made more slowly and more rapidly. A series of 15 movements was performed for each value of the inertia. The subject was instructed to relax the biceps and triceps before the test and to perform a continuous movement with respect to the amplitude of the movement.

Results and Discussion

The tension-length curve [Bouisset *et al.*, 1968] exhibits a phase where the decrease of the tension is sufficiently slight. During this phase, when the tension as well as the length decreases in an almost linear way, most of the movements show no electromyographic activity from the biceps and triceps muscles. This is indicated in figure 2 by the distance between the two arrows. On the assumption that muscular elasticity is only involved during this relaxing period where both muscles are simultaneously inactive, a

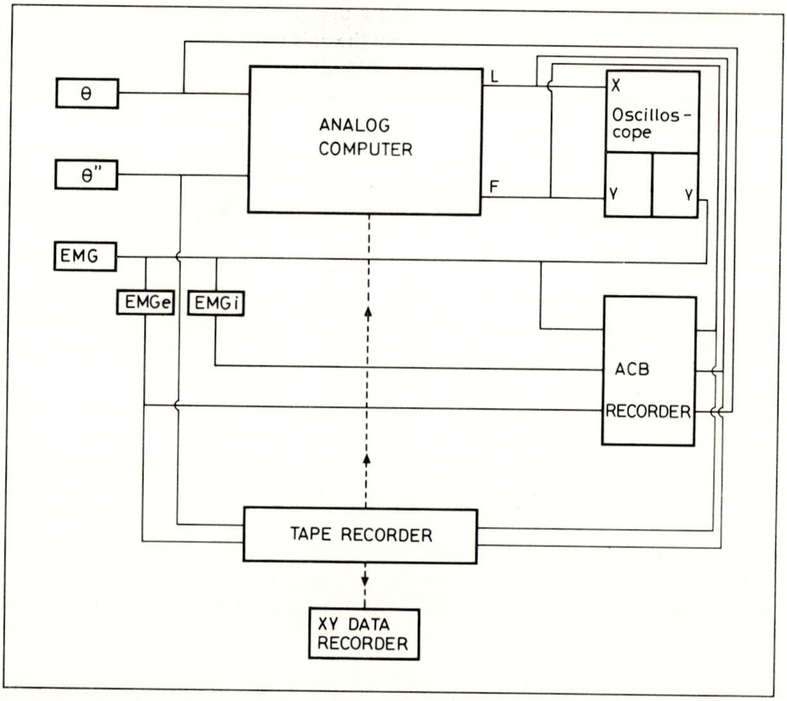

Fig. 1. Block circuit diagram.

compliance value can be calculated from the inverse of the slope of the tension-length curve ($\Delta L/\Delta F$). The investigation of the compliance variation as a function of the maximum tension developed during the movement shows that compliance increases when tension decreases (fig. 3).

The curve obtained can be interpreted in terms of HILL's two-component system. The most simple hypothesis seems to suggest that, in dynamic conditions as well as during passive extension, the tension-length curve of the elastic component is not linear, the slope of the relation ($\Delta L/\Delta F$) decreasing with tension. However, we have to point out the peculiarity of compliance values calculated in such conditions, i.e. in the course of movements performed at variable velocities. So, the relatively important values of compliance observed at lower levels of contraction may be noticed. This feature may be due to diverse factors such as the initial level of contraction, the initial length and/or the passive stretching of the antagonists or the intervention of a viscous phenomenon.

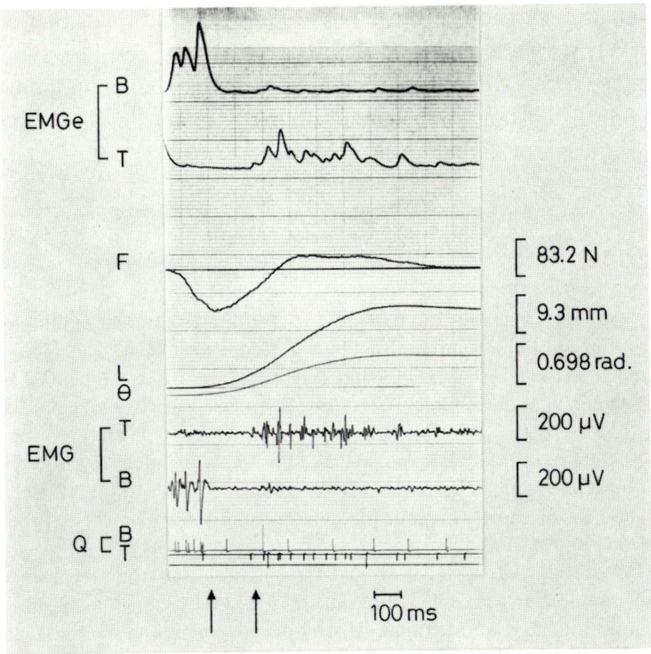

Fig. 2. Experimental curve.

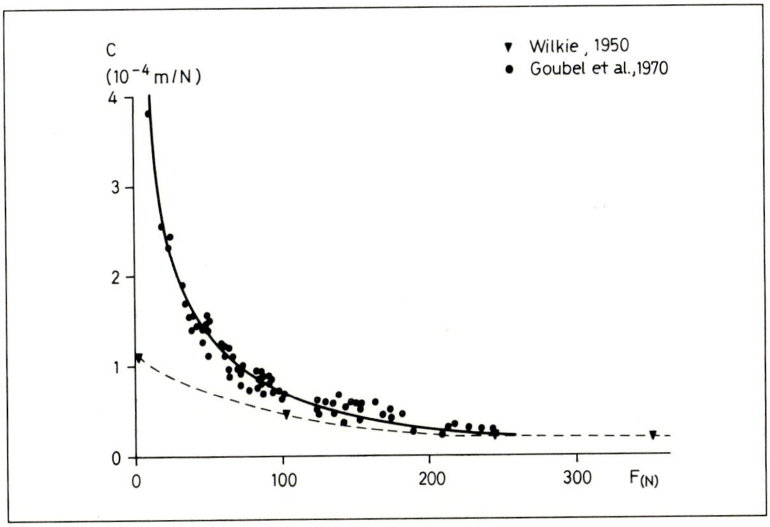

Fig. 3. Relation between compliance (C) and tension (F).

The detailed discussion of these questions is beyond the scope of this paper. However, it is interesting to emphasize that our results agree with those of WILKIE [1950] who established a relation between compliance and tension showing a constant value (about $0.3 \cdot 10^{-4}$ m/N) similar to the one shown above at least for higher levels of tension.

References

BOUISSET, S.; GOUBEL, F., and LESTIENNE, F.: Tension-length curve at variable velocity in normal human muscle. J. Physiol. Lond. *197:* 46–47 P (1968).

BOUISSET, S. and PERTUZON, E.: Experimental determination of the moment of inertia of limb segments. Biomechanics I. 1st int. Seminar, Zurich 1967, pp. 106–109 (Karger, Basel, 1968).

CAVAGNA, G. A.; DUSMAN, B., and MARGARIA, R.: Positive work done by a previously stretched muscle. J. appl. Physiol. *24:* 21–32 (1968).

HILL, A. V.: The heat of shortening and the dynamic constants of muscle. Proc. roy. Soc. B. *126:* 136–195 (1938).

MARGARIA, R.: Positive and negative work performances and their efficiencies in human locomotion. Int. Z. angew. Physiol. *25:* 339–351 (1968).

PERTUZON, E. and BOUISSET, S.: Maximal velocity of movement and maximal velocity of muscle shortening. Medicine and Sport, vol. 6, pp. 170–173 (Karger, Basel 1971).

RALSTON, H. J.; INMAN, V. T.; STRAIT, L. A., and SHAFFRATH, M. D.: Mechanics of human isolated voluntary muscle. Amer. J. Physiol. *151:* 612–620 (1947).

WILKIE, D. R.: The relation between force and velocity in human muscle. J. Physiol. Lond. *110:* 249–280 (1950).

Authors' address: Dr. F. GOUBEL, Dr. S. BOUISSET, and Dr. F. LESTIENNE, Laboratoire de Physiologie générale, Faculté des Sciences, B. P. 36, *F-59 Lille* (France)

Medicine and Sport, vol. 6: Biomechanics II, pp. 159–167
(Karger, Basel 1971)

Power Output of the Previously Stretched Muscle[1]

G. A. Cavagna, L. Komarek, G. Citterio, and R. Margaria

Istituto di Fisiologia Umana, Università di Milano, Milano

Introduction

The positive work performed by a muscle during a given shortening at a given speed on a Levin and Wyman ergometer may be appreciably greater when the shortening is preceded by stretching the contracted muscle [Cavagna *et al.*, 1968].

In the everyday life movements the muscles are often stretched before shortening: only they do not shorten necessarily at a constant speed, and the extent of their shortening is only eventually limited by the anatomy of the limbs. In addition, in voluntary contractions not all the motor units may be active simultaneously for the total period of stretching as is the case for the isolated muscle: during a maximal effort implying a reversal of the movement, the antagonistic muscles are probably progressively excited to reach full activity only in the proximity of the point of reversal (fig. 1); the force attained during stretching, in voluntary movements of this type, may be less than in the isolated muscle experiments and the effect on the positive work done by the muscle may be reduced correspondingly. Fenn *et al.* [1931] in the forward and back swing of the forearm found that the torque exerted at the turning-point was just equal to the maximum force the muscles were able to exert isometrically at that point; the fact that the force is not greater than but equal to isometric seems to indicate that not all the fibers were in activity during stretching. Furthermore, he found that the terminal velocity of the fist brought down on a foam-rubber pad on the table by extension of the elbow is the same whether the extension takes place starting from the flexed position, or whether it is preceded by a rapid flexion (personal communication); then also the kinetic energy, and the work done, must be the same and this is in apparent contradiction with the results obtained on the isolated muscle where the work done by the previously stretched muscle is greater.

On the other hand, according to Hill [1950] the mechanical energy stored in the series elastic component when the tension rises 'can be used in producing a final velocity greater than that at which the contractile component itself can shorten. This is important in such movements as jumping or throwing': the previous stretching of

1 This work has been supported by the Italian National Research Council (CNR).

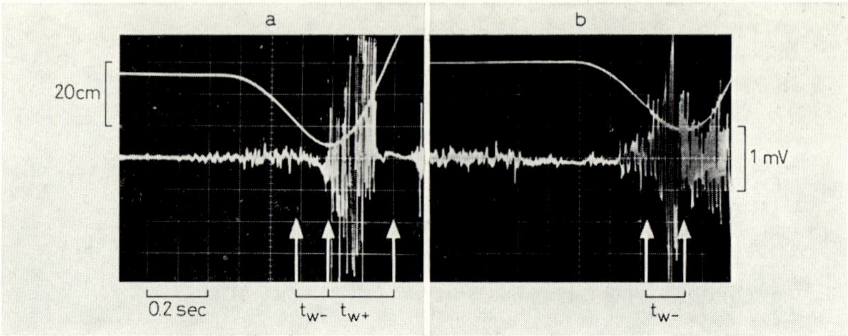

Fig. 1. Experimental tracings showing the vertical displacement of the center of gravity at the beginning of a 'normal' vertical jump (upper lines), and the EMG of the calf muscles (a) and of the extensors of the leg (quadriceps) (b). The arrows indicate the beginning and the end of the time of negative (t_W—) and of positive work (t_W+).

the contracted muscles, leading to a greater tension, should then lead to a greater final velocity.

The present experiments have been devised in order to assess the effect of a previous stretching of the contracted muscles, such as takes place in ordinary voluntary movement, on (a) the amount of positive work performed, (b) the final speed attained, and (c) the power output.

Method

The voluntary movement consisted in performing a standing vertical jump off both feet.

The subjects, 5 men and 2 women (A.P. and S.M. in table 1) aged 22–35, not particularly trained in jumping, jumped from a 50×50-cm platform (natural frequency when charged with a 68-kg subject: 116 cps) sensitive to the vertical force F. The vertical components of the velocity V, and of the displacement S of the center of gravity of the body were obtained by single, respectively double, integration of its acceleration, which was assumed to be proportional to the difference between F, the total force measured by the platform, and P, the body weight ($F - P = m\,a$). Integration was carried out by Philbric operational amplifiers. In figure 2 are shown experimental tracings of the vertical component of the force acting on the platform (F, lower tracing), of the vertical component of the speed (V, middle tracing) and of the vertical displacement of the center of gravity (S, upper tracing).

The jumps were taken with a maximal effort, starting from three different positions. In the first tracing from the left the subject performed a maximal effort starting with the legs flexed, the upward acceleration taking place without any previous muscle stretching: in this case the muscles perform only positive work. In the other

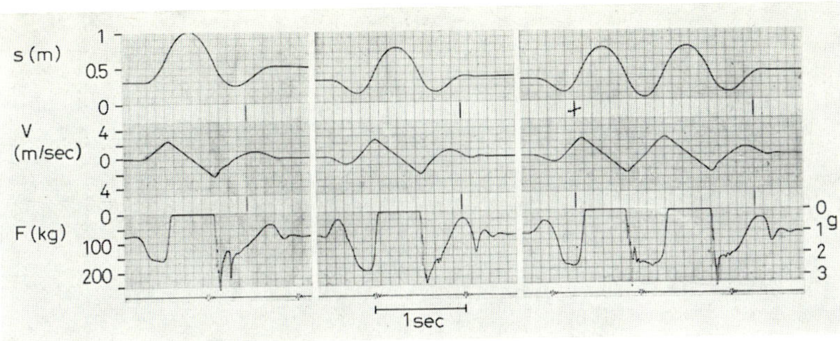

Fig. 2. The vertical component F of the resultant of the forces exerted by the feet on the ground, the vertical velocity V and displacement S of the center of gravity of the body are plotted as a function of the time, during a jump off both feet from a force platform.

two tracings the upward jump was immediately preceded by a flexion of the knees as occurs spontaneously (normal jump); this flexion is indicated by a decrease in the force F on the platform, and by a downward deflection of the V and S tracings: toward the end of the downward displacement of the body the deceleration is obtained through the breaking action of the muscles that are stretched as they absorb the potential and kinetic energy of the body (fig. 1). We also tried to emphasize the effect of such previous stretching by performing two jumps in succession as suggested by MAREY and DEMENY [1885] (last tracing on the right in fig. 2); this manoeuvre, however, did not lead to a result substantially different from that obtained with a single flexion.

The extent of flexion on the knees preceding the positive work phase was not imposed; the subjects simply chose spontaneously the extent of flexion that they felt was most efficient.

Results

When the upward movement is not preceded by stretching of the contracted muscles, the speed (V) begins to increase slowly from zero, whereas when it is preceded by stretching (downward movement), it increases at a higher rate from a negative value crossing sharply the zero line (fig. 2). In spite of this initial advantage of the previously stretched muscle, the maximal speed attained is not appreciably higher, only about 6% (table I); correspondingly the speed at the take off, v_d (as measured when the vertical force falls to zero) and, therefore, the upward displacement of the center of gravity during the flight period:

Table I. Average values of the positive work done W (kgm), the time of positive work t_w (sec), the maximal upward velocity V_{mx} (m/sec), the maximal instantaneous power $\dot{W}_{Mx}$ (HP) and the average power $\overline{\dot{W}}$ (HP), during a standing vertical jump off both feet performed immediately after stretching the contracted muscles (normal) and, without stretching, from a flexed position (flexed)

Subject	Weight	W				t_w				V_{mx}				$\dot{W}_{Mx}$				$\overline{\dot{W}}$			
		Normal		Flexed		Normal		Flexed		Normal		Flexed		Normal		Flexed		Normal		Flexed	
	kg	kgm	SEM	kgm	SEM	sec	SEM	sec	SEM	m/sec	SEM	m/sec	SEM	HP	SEM	HP	SEM	HP	SEM	HP	SEM
G.C.	76.6	57.3	1.3	59.11	0.89	0.225	0.00	0.37	0.00	2.85	0.00	2.8	0.00	5.81	0.12	5.71	0.00	3.398	0.05	2.1	0.00
A.Z.	64	57.5	0.9	44.5	2.1	0.287	0.00	0.397	0.02	2.92	0.03	2.6	0.03	4.55	0.12	4.36	0.00	2.73	0.06	1.50	0.03
A.C.	77.2	51.1	2.9	48.9	1.6	0.26	0.02	0.36	0.01	2.49	0.04	2.43	0.01	4.54	0.09	4.67	0.03	2.62	0.07	1.81	0.03
L.K.	80	52.7	0.95	49.4	0.06	0.26	0.00	0.41	0.00	2.42	0.00	2.32	0.00	4.75	0.03	4.48	0.05	2.64	0.06	1.6	0.07
T.V.	68.2	53.6	0.85	45.6	0.61	0.244	0.00	0.41	0.06	2.78	0.00	2.64	0.00	4.9	0.15	4.7	0.13	2.94	0.09	1.5	0.04
A.P.	56.1	37.6	0.8	39.2	0.9	0.236	0.00	0.435	0.00	2.55	0.00	2.36	0.03	3.44	0.03	2.98	0.1	2.12	0.00	1.2	0.03
S.M.	66.8	47.7	1.1	38.4	0.71	0.235	0.00	0.327	0.00	2.67	0.03	2.43	0.03	4.62	0.09	4.65	0.1	2.7	0.04	1.58	0.06
Mean	69.84	51.08		46.47		0.25		0.387		2.67		2.51		4.66		4.51		2.737		1.614	
SD	8.54	6.86		7.01		0.02		0.036		0.188		0.171		0.698		0.804		0.383		0.281	
SEM	3.23	2.59		2.65		0.00		0.01		0.071		0.065		0.264		0.304		0.144		0.106	

SD = standard deviation, SEM = standard error of the mean, average value of observations per subject = 6.

$S_f = v_d^2/2\,g$,

differ very little. During the flight period, indicated by $F = 0$ on the tracings, the speed decreases linearly with time at a rate of 9.8 m/sec^2, as expected, to attain a maximal negative value which does not differ appreciably from the maximal positive one: this indicates that the posture of the lower limbs at the take-off and at the landing is the same (extended). The mechanical features of the jump during most of the positive work phase are shown in detail in figure 3. This phase begins when the velocity becomes positive and ends when the vertical force falls to zero; since the point in which the speed begins to increase is not easily assessed in the jump from the flexed position, the beginning of the period of positive work was in this case taken at the point at which the force on the platform becomes greater than the body weight. The time of positive work, t_w, is about 55% greater if no previous stretching of the muscles takes place (table I).

The total external positive work done against gravity W can be measured by summing up the increases of potential and kinetic energy up to take-off, which is equal to the product of the body weight P and the total vertical displacement: $W = \Delta E_p + \Delta E_k = PS$; the total displacement is $S = S_f + S_g$, where S_g is the lift of the center of gravity when the feet are in contact with the ground. It is noteworthy that on the earth $\Delta E_p \simeq \Delta E_k$.

The positive work done immediately after stretching is only about 10% greater than the positive work done when starting from the flexed position (table I); this difference is accounted for by the slightly greater speed (and, therefore, the greater kinetic energy) at take-off, and by S_g being a little greater in the 'normal' jump.

The average power developed during the jump, $\bar{\dot{W}} = W/t_w$, is appreciably greater (70%) when the jump is preceded by the stretching (table I), whereas the instantaneous power, as obtained by the product of the force and the velocity at each instant, goes through a maximum which is not appreciably different in the two conditions (table I).

The values of peak power measured in this work are of the same order of magnitude as those reported by DAVIES and RENNIE [1968].

Discussion

The different way to perform positive work in the two conditions, with and without previous stretching, consists essentially in a difference of the early increase in the upward speed. Surprisingly, the maximal speed attained is largely the same in both cases.

The Rise of the Upward Speed

The higher rate of increase of the speed upward when the positive work phase is preceded by negative work (stretching of the contracted muscle) may be due to the fact that:

(1) the muscles are already fully active whereas, when starting from rest, a certain amount of time is required for the activity to attain the maximum value;

(2) at the end of stretching the tension of the series elastic elements of the muscle is greater, and the action of the contractile component is transmitted more efficiently.

Information about factor 1 is supplied by the work of MERTON [1954] whose tracings show that the rate of increase of the force in isometric contraction is not appreciably different during maximal artificial stimulation of the motor nerve and maximal voluntary contraction. During maximal voluntary efforts only about 40 msec are required by the respiratory muscles to attain full activity [MOGNONI *et al.,* 1968].

Factor 2 is probably more important: during stretching the contracted muscle the force rises and the series elastic elements are under higher tension at the beginning of the positive work phase. For a great part of this period the force decreases (fig. 3) and the elastic component necessarily shortens by an amount S_{el}, thus contributing with the shortening of the contractile component S_c, to the total shortening of muscle, $S_{tot} = S_c + S_{el}$. In the case of the unstretched muscles, on the contrary, for most of the positive work phase the force exerted by the contractile component increases (fig. 3), thus stretching and lenghtening the series elastic elements: the total shortening of the muscle will, therefore, be $S_{tot}=S_c-S_{el}$, and correspondingly less will be the velocity of shortening.

The Maximal Vertical Speed

The greater force developed from the beginning of the movement in the previously stretched muscles implies a greater acceleration upward and, therefore, a shorter time to reach a given speed: the fact that the maximal speed reached appears to be sensibly the same in the stretched as in the unstretched muscles, seems to indicate the existence of a limiting factor. This possibly resides in the maximal speed of shortening of the contractile component of the muscles.

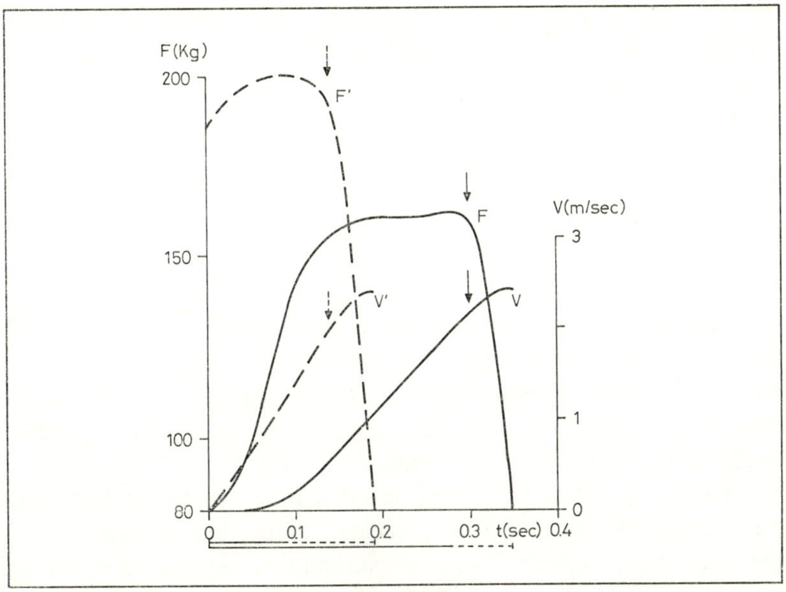

Fig. 3. Vertical force and velocity of the center of gravity during most of the phase in which the previously stretched muscles (F′ and V′, broken line) and the unstretched muscles (F and V, continuous line) perform positive work. The arrows indicate the instant in which the muscles develop the maximal instantaneous power. The horizontal lines at the bottom of the figure indicate the time during which the force rises (continuous line) and drops (broken line) during the shortening of the muscles.

When the upward speed attains its maximum, the acceleration is zero and the force supported by the limbs is equal to the body weight P; if at this time the posture of the limbs is the same in the two kinds of exercise, the force applied to the muscles, and, therefore, the speed of shortening of their contractile component v_c, will also be the same. In addition, since at this time the force falls at about the same rate in both cases (fig. 3), the speed of shortening of the series elastic elements, v_{el}, will also be about the same. It follows that also the maximal speed of shortening of the muscle as a whole $v_{tot} = v_c + v_{el}$ will also be the same in the two conditions.

The hypothesis that the speed of shortening of the contractile component is a limiting factor of the maximal upward speed is supported by the following observation: by loading the subject with a weight on the shoulders the maximal speed attained decreases with increasing weight, as expected from the force-velocity relationship of the contractile component (experiments in progress). The same hypothesis would also give reason for the maximal velocity being the same in the two conditions of the extension of the elbow in the experiment performed by FENN (personal communication).

Significance of Stretching the Contracted Muscles in Exercise

The usefulness of a movement in the opposite direction preceding a given action in many common exercises was pointed out by FENN [1930] and by HOCHMUTH [1968]; these authors noticed that at the reversal of the movement of the limbs in running, respectively in jumping, the force does not fall to zero but it has a high value from the beginning of the push.

MAREY and DEMENY [1885] stated that in two successive jumps the second is higher than the first, because of the elastic energy stored at the end of the first jump. This is not substantially confirmed by the present experiments (fig. 2), which indicate that the previous stretching does not appreciably increase the amount of positive work done. Stretching leads to greater work and power in the experiments on the isolated muscles which have been made to shorten at constant speed [CAVAGNA *et al.*, 1968]: in ordinary movements, where the speed of shortening is not necessarily constant, it only leads to a greater average power. Since the time of positive work is reduced for the previously stretched muscle, the cost for maintaining the muscle active will be less; in addition, part of the positive work performed will be due to the elastic energy stored in the series elastic elements during the stretching. In the unstretched muscle, on the contrary, all the energy responsible for the positive work performed comes from the chemical energy transformation at the contractile component level.

The lower cost of the positive work done by the previously stretched muscle may appear of little use if the cost of the preceding negative work is taken into consideration. It is a peculiar feature of most exercises, however, that negative work precedes immediately positive work, and it takes place anyway. This is particularly evident in running where negative work is performed alternately with positive work within the step. The elastic energy stored in the muscle during the negative work phase would be wasted, were it not used in the following positive work phase; it does not imply an extra energy expenditure.

References

CAVAGNA, G. A.; DUSMAN, B., and MARGARIA, R.: Positive work done by a previously stretched muscle. J. appl. Physiol. *24:* 21–32 (1968).

DAVIES, C. T. M. and RENNIE, R.: Human power output. Nature, Lond. *217:* 770–771 (1968).

FENN, W. O.: Frictional and kinetic factors in the work of sprint running. Amer. J. Physiol. *92:* 583–611 (1930).

FENN, W. O.; BRODY H., and PETRILLI, A.: The tension developed by human muscle at different velocities of shortening. Amer. J. Physiol. *97:* 1–14 (1931).

HILL, A. V.: The series elastic component of muscle. Proc. roy. Soc. B. *137:* 273–280 (1950).

HOCHMUTH, S.: Biomechanics. I. 1st int. Seminar, Zurich 1967, pp. 155–160 (Karger, Basel 1968).

MAREY, M. et DEMENY, M. G.: Locomotion humaine, mécanisme du saut. C. R. Acad. Sci. *101:* 489–494 (1885).

MERTON, P. A.: Voluntary strength and fatigue. J. Physiol., Lond. *123:* 553–564 (1954).

MOGNONI, P.; SAIBENE, F.; SANT' AMBROGIO, G., and AGOSTONI, E.: Dynamics of the maximal contraction of the respiratory muscles. Resp. Physiol. *4:* 193–202 (1968).

Authors' address: Drs. G. A. CAVAGNA, L. KOMAREK, G. CITTERIO, and R. MARGARIA, Istituto di Fisiologia Umana, Università di Milano, Via Mangiagalli 32, *Milan* (Italy)

Medicine and Sport, vol. 6: Biomechanics II, pp. 168–169
(Karger, Basel 1971)

Force and Speed in Human Muscle

K. HAINAUT

Brain Research Unit (Prof. J. E. DESMEDT), University of Brussels, Brussels

During indirect stimulation of an intact normal human muscle at a rate of 2 sec, there is a gradual increase in the twitch force, known as the staircase phenomenon [DESMEDT and HAINAUT, 1967; HAINAUT, 1968]. Details of the technique can be found elsewhere [DESMEDT, 1958]. The staircase potentiation should not be confused with the chemical potentiations such as produced by caffeine, nitrate, zinc, etc. where the tension increase is accompanied by a prolongation of the contraction time and of the relaxation time of the response [KAHN and SANDOW, 1955; SANDOW, 1964]. In staircase potentiation there is an early increase in the contraction speed of the twitch while the maximum intrinsic velocity of the muscle in tetanus is not significantly changed [DESMEDT and HAINAUT, 1968].

The purpose of this work is to show that, at any given force developed by the muscle, the speed indicated by the first derivative (Rt), is larger in the potentiated twitch. The experimental arrangement to study the instantaneous relation between the force (Pt) and its first derivative (Rt) was such that the two magnitudes were fed to the X and Y inputs, respectively, of a cathode ray oscilloscope [HAINAUT and DESMEDT, 1968]. The muscle activity makes the spot move clockwise so that it describes a loop (fig. 1) showing an upward peak that corresponds to the maximum rate of tension development, an intersection with the abscissa as Pt is maximum and a minimum on the ordinate corresponding to the maximum rate of relaxation [MCPHERSON and WILKIE, 1954]. The figure shows two superimposed loops representing the responses to the first and the 200th shock of a series of stimulations at a frequency of 2 sec.

The loop corresponding to the 200th twitch, which is potentiated, presents larger amplitudes on the abscissa and on the ordinate than the loop recorded after the first stimulation. Thus the rate of tension development to the potentiated twitch is larger than that of the first twitch at any com-

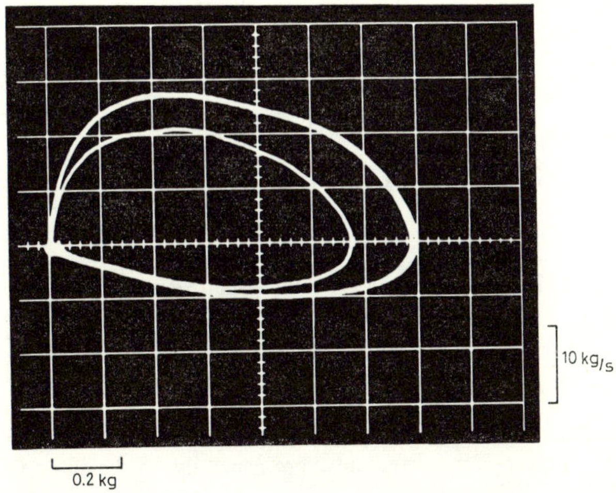

Fig. 1. Normal adult man. Isometric twitches of the adductor pollicis muscle elicited by supramaximal electric pulses to the ulnar nerve at the wrist. Cathode ray oscilloscope loops displaying muscle force (Pt) on the abscissa and the corresponding rate of tension development (Rt) on the ordinate. Loops corresponding to the first and to the 200th responses of a series at 2 sec are superimposed.

parable level of force. The staircase potentiation which is an increase in the muscle force accompanied by an acceleration of the twitch time course appears to be related to an intensification of the electro-mechanical coupling process.

References

DESMEDT, J. E.: Acta neurol. belg. *58:* 977–1017 (1958).

DESMEDT, J. E. and HAINAUT, K.: C. R. Acad. Sci. *264:* 363–366 (1967).

DESMEDT, J. E. and HAINAUT, K.: Nature, Lond. *217:* 529–532 (1968).

HAINAUT, K.: Biomechanics. I. 1st int. Seminar, Zürich 1967, pp. 255–257 (Karger, Basel 1968).

HAINAUT, K. and DESMEDT, J. E.: Arch. int. Physiol. Biochem. *76:* 554–556 (1968).

KAHN, A. J. and SANDOW, A.: Ann. N. Y. Acad. Sci. *62:* 137 (1955).

MCPHERSON, L. and WILKIE, D. R.: J. Physiol., Lond. *124:* 292–299 (1954).

SANDOW, A.: Arch. phys. Med. *45:* 62 (1964).

Author's address: Dr. K. HAINAUT, Brain Research Unit, University of Brussels, *Brussels* (Belgium)

Medicine and Sport, vol. 6: Biomechanics II, pp. 170–173
(Karger, Basel 1971)

Maximum Velocity of Movement and Maximum Velocity of Muscle Shortening

E. PERTUZON and S. BOUISSET

Laboratoire de Physiologie générale, Faculté des Sciences, Lille

The knowledge of the maximum velocity of movement is of obvious interest to the study of the performance, and more generally, to the study of the motor function and its deficiencies, if any. This is a characteristic whose value, even for a simple mono-articular movement, depends on various factors. It is known [BOUISSET, 1967] that the maximum velocity is a function of the biomechanical conditions and of the perceptual requirements of the movement considered, as well as of the intrinsic velocity of shortening of the muscles involved.

It is proposed to examine the relationship between the velocity of movement and the velocity of muscle shortening in the case of maximum flexions of the elbow. In addition, it is attempted to specify to what degree the maximum velocity of muscle shortening determined *in situ* can correspond to the intrinsic velocity, i.e. the velocity of shortening of the disconnected muscle in a state of maximum contraction.

Technique and Procedure

The subjects are seated. They perform flexions of the right elbow at maximal speed, successively in a vertical and in a horizontal plane.

In the first case, the arm is fixed in a vertical position and the forearm, movable in a vertical plane, is connected to a hinged system which is regulated to reproduce the movement and to control a goniometer whose axis of rotation coincides with that of the elbow. The hand is in a semi-prone position with the wrist immobilized by a splint made of plastic material. The moment of inertia of this mechanical system is 0.0035 kg · m². In the second case, the mechanical system is identical, but the arm is fixed in a horizontal position and the forearm is attached to a splint which is movable around a vertical axis coinciding with that of the elbow. The moment of inertia of this mechanical system is 0.033 kg · m².

In addition to the surface EMGs of the biceps brachii, brachioradialis and triceps brachii muscles, the angular displacement and velocity of movement as well as the velocity of shortening of biceps brachii, are being continuously recorded. The angular displacement of the forearm (Θ), calculated in relation to the position of complete extension, is measured by a goniometer. The angular velocity (Θ') is obtained by differentiation of the goniometric signal. The instantaneous value of the velocity of shortening of biceps brachii is calculated by means of an on-line analogue computer; this velocity (u) is given by the relationship shown in figure 1. The constants a and b represent the distances of insertions CB and CA of biceps brachii with respect to the axis of rotation of the elbow. If d is the distance between the styloid apophysis and the epithrochlea-epicondyle axis, the constants a and b can be calculated for each subject through the relationships

$$a = 0.18\ d \quad \text{and} \quad b = 1.17\ d,$$

where the numerical coefficients have been determined from the measurements taken from a skeleton.

Six subjects were examined twice at 24-hour intervals. They were asked to perform movements at maximum velocity and, by exerting as much force as possible, to strike out the forearm as far as the security check placed at the end of flexion. These security checks limit the amplitude to 2.511 rad (s = 0.175) in the vertical plane and to 2.288 rad (s = 0.035) in the horizontal plane. During each examination, five flexions were performed in the horizontal plane and five in the vertical plane. For each subject, only the highest values obtained in each plane were taken into account.

Results and Discussion

As shown in table I, the average of the maximum angular velocity and that of the maximum muscular velocity are higher for the horizontal movement than for the vertical movement. Although not statistically significant, this result can in all likelihood be explained by the action of gravity. On the other hand, it is remarkable that the maximum angular velocities should hardly be different from those obtained in the course of a previous experiment performed with the same subjects [BOUISSET *et al.*, 1966] and also from those indicated by DERN *et al.* [1947]. Also to be noticed is the higher

Table I. Individual values, averages and standard deviations of maximum angular (Θ') and muscular (u) velocities

Subject	Vertical		Horizontal	
	Θ' (rad/s)	u (cm/s)	Θ' (rad/s)	u (cm/s)
F.B.	15.38	87.88	16.48	74.93
S.B.	16.73	85.66	17.43	78.63
J.D.	18.04	77.70	18.04	85.10
F.G.	18.45	60.68	18.37	79.55
F.L.	15.38	72.52	18.78	85.10
E.P.	16.52	69.75	18.66	85.10
m	16.75	75.70	17.96	81.40
s	1.09	10.22	0.89	4.33

value of the variance of muscular velocity for the vertical movement. In this context, one cannot exclude the possibility that the relative contribution of the various flexors may differ for a given subject depending on whether the movement is performed in the presence or in the absence of gravity. Considering these remarks, the results obtained for the horizontal movement are essentially the ones that will be considered later.

The velocity of shortening is calculated according to the formula in figure 1. It appears as a function of the angular velocity and displacement, with the biomechanical constants *a* and *b* taken into account. One of the ways to evaluate the relation between the maximal values of muscular and angular velocities can consist in the calculation of the correlation between these two variables (fig. 2). The very high value of the coefficient of correlation ($r = .96$) seems to be the result from the good reproducibility of the movement, i.e. from the reproducibility of the muscular patterns, as suggested above. In addition, it is worth noting that the slope of the regression line seems linked to the dephasing between the angular velocity and the velocity of the biceps shortening which may be seen from the experimental curves. In any case, the relation permits an estimate of the maximum velocity of biceps shortening from the maximum velocity of flexion of the forearm with a precision acceptable according to the standard deviation of the marginal distribution ($s_{y,x} = 2.23$).

The average value of the velocity of muscular shortening calculated over six subjects is less than the one (87 cm/s) determined according to the data of WILKIE [1950] and less than the one (92.5 cm/s) determined by RALSTON *et al.* [1949] in their consideration of disconnected muscles. These differences show the importance of the conditions under which the movement is performed. This is particularly relevant in the case of inertia, for which in this study no correction allowance is made with the effect of limiting the maximum value.

On the other hand, the relatively lower values of the results here obtained can be attributed neither to the amplitude of the movement, which is not far from maximum amplitude, nor to the possible activity of antagonistic muscles [see PERTUZON and LESTIENNE, 1968]. It is finally appropriate to emphasize that the precision with regard to muscular velocity depends on the constants *a* and *b* appearing in the equation referred to: a variation of 5 mm (i.e. 10%) for *a* and of 10 mm (i.e. 3%) for *b* are corresponding to an error less than 9.7%.

It thus seems possible, provided one can evaluate the effect of inertia, to determine the intrinsic velocity of biceps shortening *in situ*.

1, 2

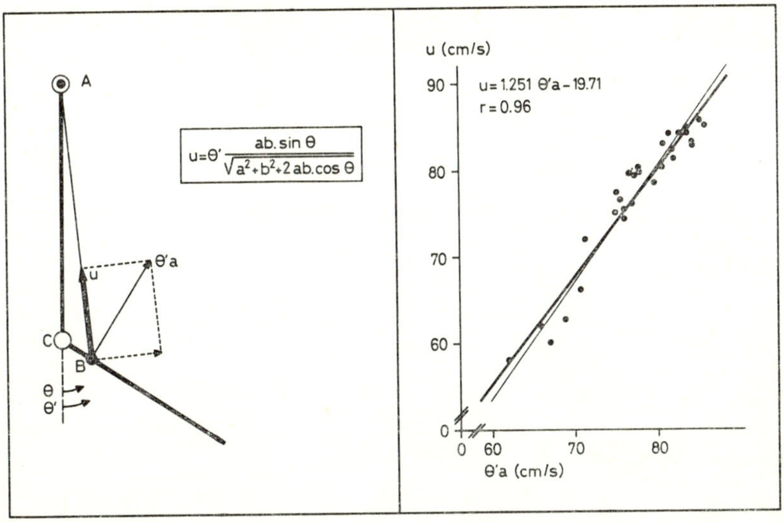

Fig. 1. Calculation of the velocity of muscular shortening. Explanations in the text.

Fig. 2. Relation between the maximum velocity of biceps shortening (u) and the maximum tangential velocity of the point of insertion of biceps with the forearm *(Θ'a)*.

The two regression lines are represented. The heavy line corresponds to the equation $u = f(\Theta' a)$. The 30 experimental values involving the horizontal movement have been taken into consideration.

References

BOUISSET, S.: Postures et mouvements; in SCHERRER Physiologie du travail, vol. 1, pp. 43–153 (Masson, Paris 1967).

BOUISSET, S.; CNOCKAERT, J.C., and PERTUZON, E.: Sur la vitesse maximale de raccourcissement du muscle au cours d'un mouvement monoarticulaire simple. J. Physiol., Paris *58:* 474 (1966).

DERN, R.J.; LEVENE, J.M., and BLAIR, H.A.: Forces exerted at different velocities in human arm movements. Amer. J. Physiol. *151:* 415–437 (1947).

PERTUZON, E. and LESTIENNE, F.: Caractères électromyographiques d'un mouvement monoarticulaire exécuté à vitesse maximale. J. Physiol., Paris *60:* suppl. 2, p. 513 (1968).

RALSTON, H.J.; POLISSAR, M.J.; INMAN, V.T.; CLOSE, J.R., and FEINSTEIN, B.: Dynamic features of human isolated voluntary muscle in isometric and free contractions. J. appl. Physiol. *1:* 526–533 (1949).

WILKIE, D.R.: The relation between force and velocity in human muscle. J. Physiol., London, *110:* 249–280 (1950).

Authors' address: Dr. E. PERTUZON and Dr. S. BOUISSET, Laboratoire de Physiologie générale, Université des Sciences et Techniques, B.P. 36, *F-59 Villeneuve d'Ascq* (France)

Medicine and Sport, vol. 6: Biomechanics II, pp. 174–180
(Karger, Basel 1971)

Maximum Strength of Elbow Flexors with Pronated and Supinated Forearm

K. JØRGENSEN and S. BANKOV

The Danish National Association for Infantile Paralysis, Hellerup

Introduction

The purpose of this investigation was to measure the maximum torque of the elbow flexors under static and dynamic conditions when the forearm was either supinated or pronated.

Method

The instrument used consisted of a lever arm system which could be moved by a reversible electric motor. Between the motor and the lever arm system a gear reduction box and a speed regulator were inserted. The degree of flexion of the elbow joint was recorded electrically for every 15°.

The axis of the lever arm was in line with the axis of the elbow joint, and the forearm was always in parallel with the lever arm. The subject's forearm was connected to the end of the lever arm by means of a cuff around the wrist. A ring-shaped strain gauge dynamometer was inserted between the cuff and the lever arm. The torque of the elbow flexors is the force measured by the dynamometer multiplied by the distance from the elbow joint axis to the cuff.

The electromyogram (EMG) from m. biceps was integrated by a Disa mean voltage integrator to give the integrated EMG (iEMG). All measurements were recorded simultaneously on photographic paper by the Disa electromyograph. The mean peak voltage of the iEMG from m. brachioradialis was estimated by a graphical method [MØLLER, 1958].

The data of the three male subjects are shown in table I.

Procedure

The subject was sitting by the side of the apparatus with his right arm placed in the lever arm system. The upper arm was in a vertical position and the forearm

Table I. The data of the subjects

Subjects	Age years	Body height cm	Forearm length (elbow to wrist) cm	Maximum isometric strength of elbow flexors in kp×cm torque, elbow angle 90°, forearm sup.
S. B.	41	176.0	26.5	700
I. N.	37	192.0	30.0	712
K. J.	30	176.5	25.6	724

supinated. In accordance with Basmajian and Latif [1957] the elbow angles are given as the compliment of an angle between the forearm and the upper arm. For the isometric measurements an elbow angle of 90° was used. In this position the maximum isometric torque of the elbow flexors was measured twice. The same measurement was repeated with the forearm pronated. Afterwards the arm was placed with an elbow angle of either 45 or 135°. In these positions the subject performed a maximum isometric contraction, immediately followed by a maximum dynamic muscle contraction, concentric, if the start position was 45°, eccentric if the start position was 135°. The speed of the contraction was set at a constant rate by the motor. The following velocities were used: 0.55, 1.07, and 2.31 rad/sec. Four experiments with each velocity on each subject were performed. In each experiment two maximum concentric and two maximum eccentric contractions were performed with the forearm supinated or pronated. As an expression of the force mobilized in the dynamic contractions the strength recorded when the elbow angle passed 90° was used.

Results

All the measurements, both torques and EMG, were expressed as a percentage of the maximum isometric values at an elbow angle of 90° and the forearm supinated.

Isometric Contractions

Figure 1 gives the relative values of maximum torque and iEMG of m. biceps brachii and m. brachioradialis at three different elbow angles with the forearm pronated or supinated. We found that the strength of the elbow flexors with the forearm pronated was about 85% of the strength obtained with supinated forearm. This is in good agreement with the results of Wells [1950]. Further we found that the maximum isometric torques at elbow angles 45 and 135° were only about 75% of the value obtained at

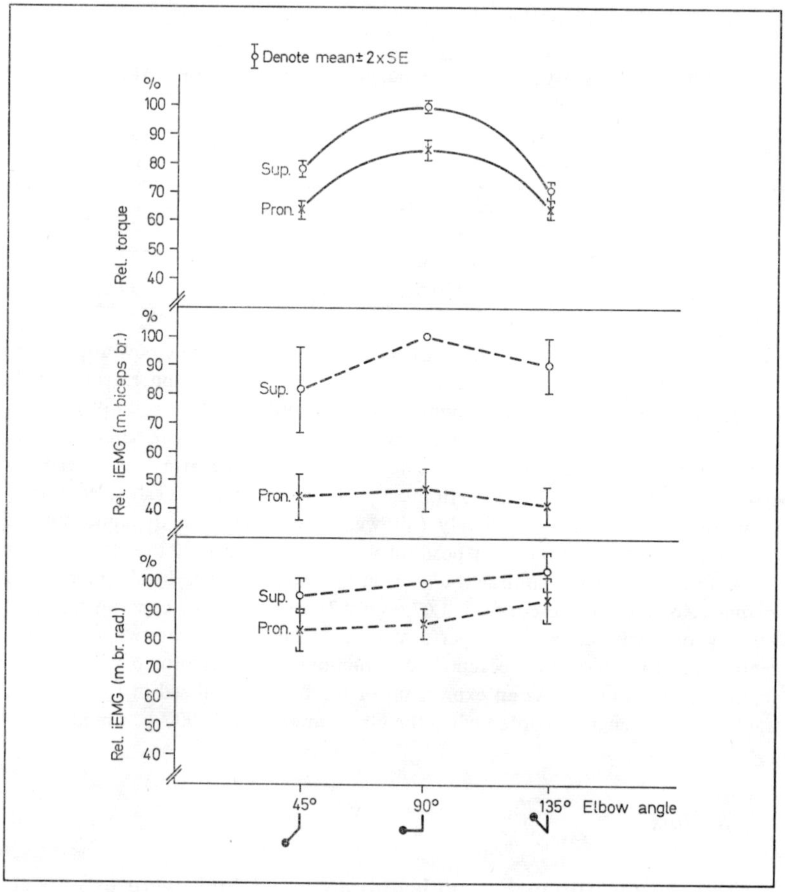

Fig. 1. Mean values of relative maximum isometric torques in elbow flexion and relative iEMG from m. biceps br. and m. brachioradialis at different elbow angles with the forearm supinated and pronated. 100% represents values obtained with supinated forearm and an elbow angle of 90°.

an elbow angle of 90°. This is in good agreement with the findings of ROHMERT [1962] and of KARPOVICH *et al.* [1966]. In m. biceps we found no significant differences in the EMG at the different elbow angles investigated, when the forearm was either supinated or pronated. But when the forearm was pronated the iEMG from m. biceps was only about 50% of that obtained when the forearm was supinated. In m. brachioradialis no significant difference in the iEMG was found with different elbow angles or as a result of the supination or pronation of the forearm.

Dynamic Contractions

During dynamic contractions at various speeds (fig. 2) curves similar to the force velocity curves from earlier investigations [WILKIE, 1950; ABBOTT *et al.*, 1952; and ASMUSSEN *et al.*, 1965] were obtained, both with the forearm in supination and in pronation. The curve from the experiments with pronated forearm was, however, 10–15% lower than the one from experiments with supinated forearm. Under dynamic conditions the electrical activity in m. biceps was 30–40% less with the forearm pronated than with the forearm supinated.

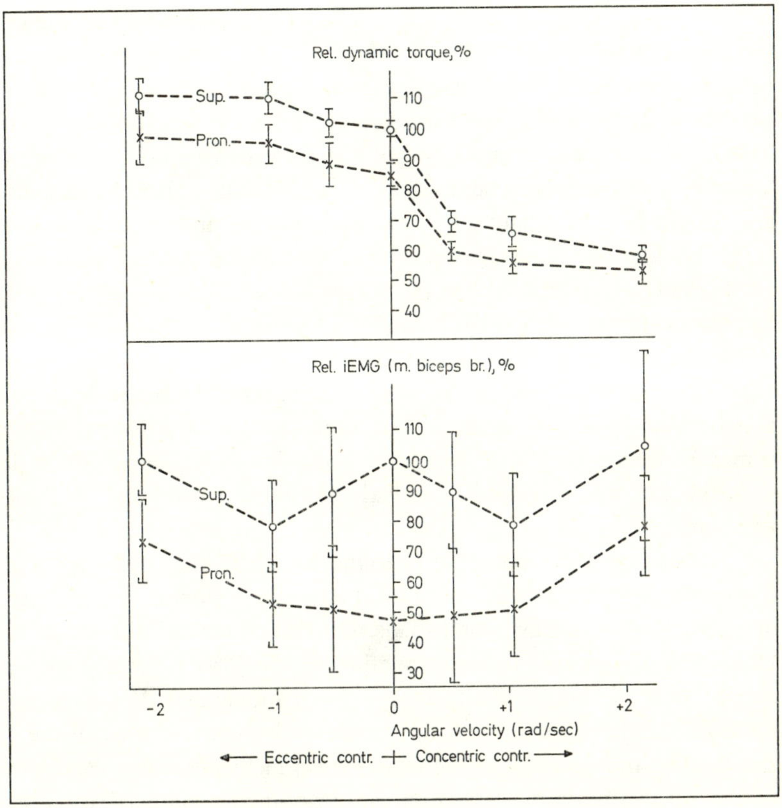

Fig. 2. Mean values of relative maximum dynamic torques of elbow flexors and relative iEMG from m. biceps br. at an elbow angle of 90° at different positive and negative velocities. 100% represents maximum isometric torque.

The iEMG from m. brachioradialis scattered very much. No systematic difference in relation to the position of the forearm (pron. or sup.) could be seen in the iEMG of this muscle.

Discussion

The present experiments have shown that the maximum isometric torque of the elbow flexors depends on the elbow angle and on the position of the forearm (fig. 1). The differences between the maximum isometric torques of the elbow flexors with the forearm pronated or supinated decreased when the elbow angle approached 135°, probably because of contact between the muscle bellies of the fore- and upper arm. In concentric contractions (fig. 2) both curves (sup. and pron.) are more or less similar to force velocity curves from other muscle groups *in situ* [cf. WILKIE, 1950]. The relative force measured during eccentric contractions was, however, less than that found by ASMUSSEN *et al.* [1965]. The reason for this discrepancy might be that ASMUSSEN *et al.* used another and larger muscle group (arm-shoulder muscles). Our results do not agree with the predictions of FICK [1911]. He estimated that the performance of the elbow flexors should be better with the forearm pronated than supinated. If the innervation of all elbow flexors were independent of the degree of supination/pronation FICK's predictions would probably have been right. M. biceps br., however, acts both as an elbow flexor and as supinator of the forearm. Therefore, it may be assumed, that m. biceps, in a maximum isometric elbow flexion with the forearm pronated is reflexly inhibited (reciprocal inhibition of antagonist).

BASMAJIAN and LATIF [1957] found by EMG that m. biceps with the forearm pronated plays only a minor if any role in flexion and maintenance of elbow flexion against small loads. Our investigations show that the inhibition of biceps due to the pronation of the forearm is also present at maximum loads and at different speeds of movement. We have tried to compare the decrease in the total maximum isometric torque of the elbow flexors with the decrease of the iEMG from m. biceps during extreme pronation of the forearm. Based on anthropometric data of BRAUNE and FISCHER [1890] (table II) the lever arms of the elbow flexors, with regard to the elbow axis, and the cross-sectional areas of the different muscles were computed for an elbow angle of 90° and with the forearm supinated.

Under the assumption that the innervation of the elbow flexors is maximal in our experiments with the forearm supinated, the total sum of the products: lever arm times cross-sectional area, should give an arbitrary expression of the maximum isometric torque of the muscle group. The results of this calculation are shown in table II. It can be seen that 34% of the total maximum isometric torque should originate from m. biceps. We found that a pronation decreased the iEMG of the biceps to 47% at an elbow angle of 90°, which means that the percentage torque of m. biceps with the forearm supinated is now reduced to $34 \times 47/100\% = 16\%$ of the total torque. This latter, therefore, should be reduced to $100-(34-16)\% = 82\%$ of the maximum isometric torque with the forearm supinated. This estimate is in good agreement with our experimental findings of an average decrease to 85%.

Table II. Estimated torque of the elbow flexors

	(1) Cross-section cm^2	(2) Lever arm cm	(3) Torque (1)×(2)	Relative torque of the individual muscles %
Pronator teres	2.7	1.34	3.62	4
Ext. carp. rad. longus	3.0	3.46	10.38	12
Brachialis	8.7	3.22	28.01	32
Biceps brachii	6.7	4.44	29.75	34
Brachioradialis	2.1	7.93	16.65	19
			88.41	

References

ABBOTT, B.C.; BIGLAND, B., and RITCHIE, J.M.: The physiological cost of negative work. J. Physiol., Lond. *117:* 380 (1952).

ASMUSSEN, E.; HANSEN, O., and LAMMERT, O.: The relation between isometric and dynamic muscle strength in man. Comm. Dan. nat. Ass. for infant. Paralys., No. 20 (1965).

BASMAJIAN, J.V. and LATIF, A.: Integrated actions and functions of the chief flexors of the elbow. J. Bone Jt Surg. *39A:* 1106 (1957).

BRAUNE, W. und FISCHER, O.: Die Rotationsmomente der Beugemuskeln am Ellbogengelenk des Menschen. Abh. sächs. Ges. (Akad.) Wiss. *15:* 242 (1890).

FICK, K.: Handbuch der Anatomie und Mechanik der Gelenke, vol. 3, p. 320 (Jena 1911).

KARPOVICH, P.V. and SINGH, M.: Isotonic and isometric forces of forearm flexors and extensors. J. appl. Physiol. *21:* 1435 (1966).

MØLLER, E.: Methodological investigations concerning electromyographic analysis of muscle coordination. Trans. europ. Orthod. Soc., p. 328 (1958).

ROHMERT, W.: Muskelarbeit und Muskeltraining. Arbeitsmed. Soz. med. Arbeitshyg. *22:* 122 (1968).

WELLS, K.: Kinesiology; 1st ed., p. 222 (Philadelphia/London 1950).

WILKIE, D.R.: The relation between force and velocity in human muscle. J. Physiol., Lond. *110:* 249 (1950).

Authors' address: Dr. KURT JØRGENSEN and Dr. STEFAN BANKOV, The Danish National Association for Infantile Paralysis, Tuborgvej 5, *DK 2900 Hellerup* (Denmark)

Medicine and Sport, vol. 6: Biomechanics II, pp. 181–184
(Karger, Basel 1971)

A Method of Measuring Some Kinetic Properties of Voluntary Muscle Activity

S. OTÁHAL

FTVS-Charles University, Department of Biomechanics, Prague

Introduction

To understand human movement it is necessary to know the values of all co-operating forces during actual muscle activity. Transforming the forces into moments with respect to the joint for an isometrical contraction, we may write the general formula:

$$\overline{M}_A + \overline{M}_{AN} + \overline{M}_P + \overline{M}_L = 0,$$

where $\overline{M}_A$ indicates the vector of the moment of active agonists, $\overline{M}_{AN}$ the vector of the moment of active antagonists, $\overline{M}_P$ the vector of the moment of passive resistances which are due to passive lengthening of ligaments and parts of muscles and friction in the joint, and $\overline{M}_L$ the vector of the movement of load, acting from outside on the extremity.

Method

The method used for the determination of the passive resistance (M_P) of a knee joint is shown in figures 1 and 2. The subject reclines on his side in a horizontal position on a table. The shin of his leg is fastened to the plate which rotates in the horizontal plane and is driven by the measuring frame. The measuring frame comprises one pair of tensometrical transducers by which the external driving force is transmitted. During the measurements the extremity which must be relaxed is moved over the whole range at a constant angle velocity of 10 °/sec. By monitoring the electromyogram (EMG) of the muscles concerned the relaxed condition is continuously watched. The output signal of the tensometrical transducers is amplified by the tensometrical bridge and visualized by an oscilloscope and a recorder. Simultaneously the position of the extremity is recorded by an electric goniometer. The absolute error of measurement of the apparatus is ± 0.02 kgm.

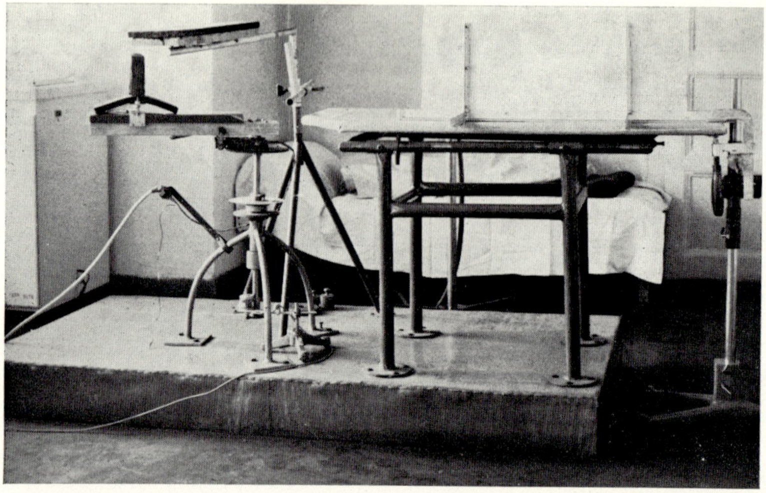

Fig. 1. Apparatus for measuring the static passive resistance in the knee joint.

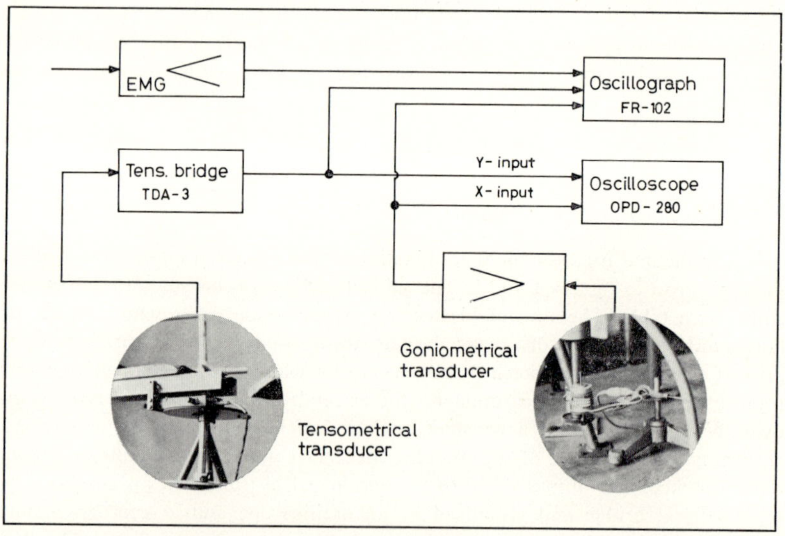

Fig. 2. Block diagram of the measuring method.

In figure 3 an example of a record of a complete measurement is shown (from the starting position to maximum extension, from maximum extension to maximum flexion and from maximum flexion backwards to the starting position.

In this way 26 persons were measured. The average range of the moment M_p between the angle 60–195 ° was approximately from –1.6 to +1.4 kgm.

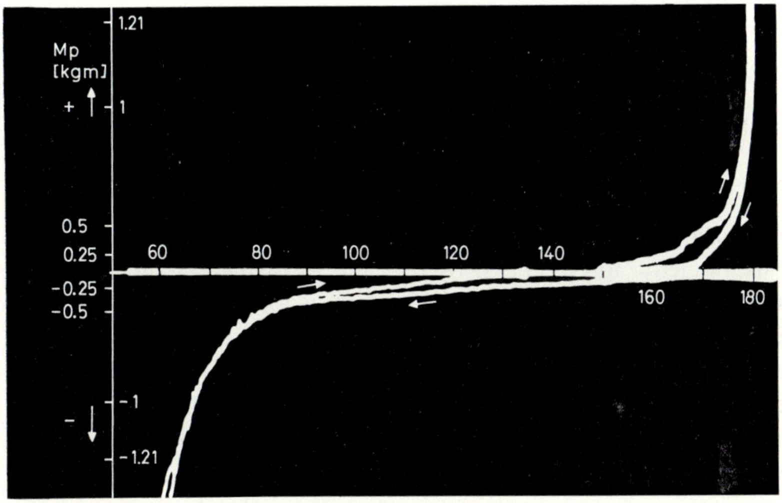

Fig. 3. Record of the values of the passive resistances during one measuring cycle.

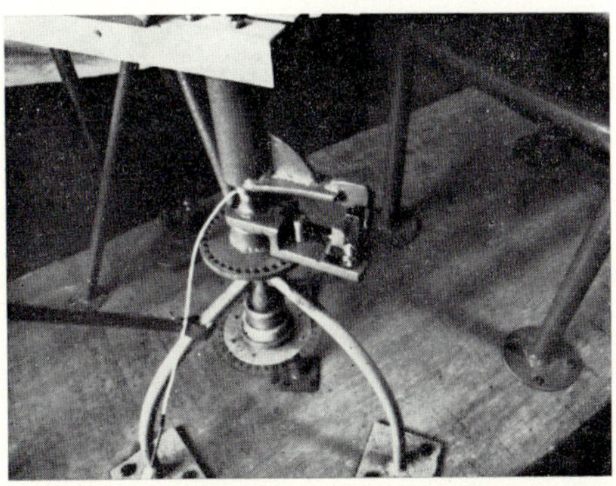

Fig. 4. Tensometrical block for measuring the isometrical forces.

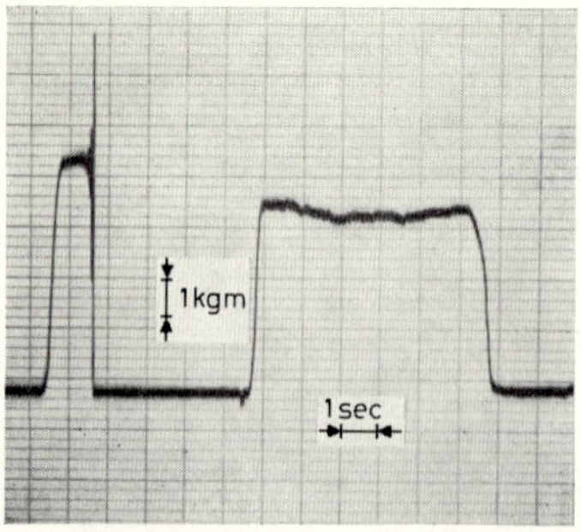

Fig. 5. Recording of two types of build-up of an isometrical force.

By simple adaptation of the device by supplying the frame with a special measuring block (fig. 4), it was possible to measure maximum voluntary isometric force exerted by the muscles of the extremity. By this method we determine the range of isometrical forces of various muscles. An example of recording is shown in figure 5. The first recording is the isometric impact, the second is the maximum voluntary contraction exerted during 5 sec.

The next stage of the experimental part of our research will be the measurement of the force-velocity characteristics of the muscles of the leg, the measurement of invariant characteristics according to Asatrjan and Feldman [1965] and the dynamic aspects of the EMG activity. These measurements will be made with the aid of the adapted apparatus which has been described.

Reference

Asatrjan, D.G. and Feldman, A.G.: O funkcionalnoj nastrojke sistemy upravlenii dviženiem ili sochranenii stacionarnoj pozy. Biofizika, Moscow *10:* 837 (1965).

Author's address: Dr. Stanislav Otáhal, FTVS-Charles University, Department of Biomechanics, Újezd 450, *Prague 1* (Czechoslovakia)

Medicine and Sport, vol. 6: Biomechanics II, pp. 185–188
(Karger, Basel 1971)

Electromyography of the Lumbar Part of the Erector Spinae Muscle

B. Jonsson

Department of Anatomy, University of Göteborg, Göteborg

Introduction

Low back pain is a common disease entity with different clinical subsyndromes. In these cases the function of the lumbar part of the erector spinae muscle is often regarded as abnormal despite the fact that the normal function is not fully known. The function of the erector spinae muscle has been the subject of several electromyographic studies, one of the first being published by Allen in 1948. Only a few authors [Morris *et al.*, 1962; Pauly, 1966; Pauly and Steele, 1966] have made more comprehensive studies on the function of different muscles in the erector spinae muscle. They used wire electrodes which were inserted into the muscles after palpation of relevant skeletal structures. These authors were not able to check the actual position of the electrodes, however.

In order to analyse the function of different muscles in the lumbar part of the erector spinae muscle these muscles have to be examined at different lumbar levels. The insertion of the electrodes may be guided by fluoroscopy, after which the location of the electrodes can be checked by injecting a gaseous contrast medium, carbon dioxide, through the EMG electrode or the insertion needle when wire electrodes are used [Reichmann and Jonsson, 1967]. The carbon dioxide stays inside the muscle fascia and can be visualized radiologically. A few minutes after the injection of the carbon dioxide the EMG activity returns to normal.

The type of EMG electrodes which seems to be best suited for electromyographic kinesiology and which have been used in this investigation are wire electrodes. Due to the fact that migration of the wires for one or more millimetres is very common [Jonsson and Reichmann, 1969] and to the fact that the pick-up area of wire electrodes probably has a diameter of several millimetres, it has been assumed that it is not possible to study any other muscles in the lumbar part of the erector spinae muscle than the multifidi, the longissimus and the iliocostalis muscles. The other muscles are considered to be too small to be studied with present electromyographic techniques.

The aim of the present study was to apply and test the method for insertion of EMG electrodes referred to above and to make a preliminary study of the function of the lumbar parts of the multifidi, the longissimus and the iliocostalis muscles in a few standing and sitting postures.

Material

The investigation was performed on 14 healthy subjects (5 males and 9 females) aged 21–27 years. A total number of 82 pairs of wire electrodes were inserted at different levels between the Th 11 and S 1 levels. In 71 of these cases the location of the electrodes could be determined radiologically, and in 56 of these cases EMG were obtained (22 from the multifidi, 29 from the longissimus, and 5 from the iliocostalis).

Methods

The hypodermic needle with which the wire electrodes were inserted was extended with a metal tube attached to a syringe for the injection of carbon dioxide. The hypodermic needle was coated with varnish except for the tip and could thus be used as a unipolar needle electrode during the insertion. The insertion of the electrodes was followed by TV-fluoroscopy with antero-posterior direction of the X-rays. Not until the radiographic examination indicated a suitable position and the EMG activity recorded with the insertion needle indicated an intramuscular location was the insertion stopped. Before the needle was withdrawn leaving the wire electrode ends in the muscle, usually between 10 and 20 ml of carbon dioxide were slowly injected.

The subsequent electromyograms were obtained with a standard electromyograph, type Disa. Only static contractions in different postures were studied. In the standing posture, standing at ease and standing with loads of 10 kp in one or both hands were investigated. Resisted flexion, extension, lateral flexion and rotation were examined in the sitting posture.

Radiographic Location of the EMG Electrodes

The carbon dioxide was usually easy to detect on the fluoroscopic screen and in the radiograms. The gas always spread along the muscle fibres, usually for a distance of several vertebral segments. The identification of level was, therefore, based on the location of the tip of the insertion needle and the identification of muscle was based on the distribution of the carbon dioxide and the direction of the gas streaks.

Most of the carbon dioxide injected into the multifidi muscles was found to be located along the lateral border of the transversospinalis compartment. The gas streaks had a cranio-medial direction towards the spinous processes.

Typical of the longissimus muscle was a longitudinal arrangement of the gas streaks in the antero-posterior projection. In the medial part of the muscle these streaks had a slight cranio-medial inclination in the upper lumbar spine and a pronounced cranio-medial inclination in the lower lumbar spine. The gas in the longissimus muscle did not or only slightly reach medial to the lateral border of the multifidi muscles.

When carbon dioxide was injected into the iliocostalis muscle, the gas was located laterally in the erector spinae muscle and the gas streaks always had a cranio-lateral direction towards the lower ribs.

Function of the Erector Spinae Muscle

In the standing-at-ease position the multifidi muscles were usually slightly active in all lumbar levels, while the longissimus muscle was most active in the L 3 level and usually not active in the other lumbar levels. The iliocostalis muscle was usually not active in standing at ease. When present, the activity was either constant or showed slight variations in intensity.

Approximately the same result was obtained in standing with both hands loaded with 10 kp. When a load of 10 kp was held in one hand no activity at all was recorded in the erector spinae muscle on the same side as the load was held, while on the unloaded side moderate activity was recorded from the iliocostalis muscle and moderate or slight activity from the whole lumbar part of the longissimus muscle. The multifidi muscles on the unloaded side did not show any obvious increase in activity. When a load of 10 kp was held with both hands in front of the trunk all three muscles showed increase in activity.

Resisted flexion of the trunk in the sitting position usually caused a relaxation of the multifidi and the longissimus muscles, while resisted extension of the trunk gave an increase in activity in all three muscles. Lateral flexion of the trunk did usually only slightly increase the activity from the multifidi muscles, while the activity from the ipsilateral longissimus and iliocostalis muscles increased markedly. Rotation of the trunk gave an obvious increase in activity in the ipsilateral iliocostalis muscle and in the contralateral multifidi muscles and the contralateral lower lumbar parts of the longissimus muscle.

Concluding Remarks

The results obtained in this preliminary study indicate that there exists a functional difference between the multifidi, the longissimus and the iliocostalis muscles in the lumbar spine as well as between different lumbar levels in these muscles. It, therefore, seems justifiable to continue the study of the function of the lumbar part of the erector spinae muscle as well as other parts of the erector spinae muscle.

References

ALLEN, C.E.L.: Muscle action potentials used in the study of dynamic anatomy. Brit. J. phys. Med. *11:* 66–73 (1948).

JONSSON, B. and REICHMANN, S.: Displacement and deformation of wire electrodes in electromyography. A roentgenologic study. Electromyography *9:* 201–211 (1969).

MORRIS, J.M.; BENNER, G., and LUCAS, D.B.: An electromyographic study of the intrinsic muscles of the back in man. J. Anat. *96:* 509–520 (1962).

PAULY, J.E.: An electromyographic analysis of certain movements and exercises. I. Some deep muscles of the back. Anat. Rec. *155:* 223–234 (1966).

PAULY, J.E. and STEELE, R.W.: Electromyographic analysis of back exercises for paraplegic patients. Arch. phys. Med. *47:* 730–736 (1966).

REICHMANN, S. and JONSSON, B.: Contrast radiography with carbon dioxide for the location of intramuscular EMG electrodes. Electromyography *7:* 103–124 (1967).

Author's address: Dr. BENGT JONSSON, Department of Anatomy, University of Göteborg, *Göteborg* (Sweden)

b) Principles of Motion

Medicine and Sport, vol. 6: Biomechanics II, pp. 190–191
(Karger, Basel 1971)

Telemetering Electromyographic Studies of Gait in Man

J. JOSEPH

Department of Anatomy, Guy's Hospital Medical School, London

It is possible to dispense with long trailing wires and/or a treadmill by using a telemetering electromyograph. Surface electrodes are used. 8 men and 6 women were investigated while walking on a flat surface. The main observations are as follows. The tibialis anterior is markedly contracted at the beginning and end of the swing phase but during the swing phase usually shows absence of or reduced contraction. It also contracts in the early part of the stance phase. The soleus contracts during the stance phase but its action begins and ends earlier than expected. The quadriceps femoris contracts at the end of the swing phase and beginning of the stance phase, and also in some subjects at the end of the stance phase. The hamstrings contract during the second half of the swing phase and early part of the stance phase. In none of the subjects do they contract during the middle part of the swing phase. The hip flexors contract during the first half of the swing phase. The gluteus medius contracts during the stance phase and in some subjects during the swing phase. The gluteus maximus contracts during the early part of the stance phase. Both erector spinae muscles contract twice, in the early stance and early swing phases.

When walking upstairs, the body is lifted on to the next stair by the soleus, quadriceps femoris, hamstrings and gluteus maximus and the gluteus medius prevents the body falling to the unsupported side. The tibialis anterior contracts only during the middle of the swing phase. Both erector spinae muscles contract twice, once in the early part of the swing phase and again in the late part of the swing and early part of the stance phase. This, however, is variable.

When walking downstairs the body is lowered by the controlled lengthening of the soleus and quadriceps femoris. The tibialis anterior contracts twice, once in the early swing phase and once in the early stance phase. Both erector spinae muscles contract twice, once during the early swing phase and once during the early stance phase.

When walking on high heels, women show comparatively little change in the activity of the muscles investigated. The tibialis anterior contracts continuously during the stance phase, but much less markedly than when walking on low heels. The soleus contracts more strongly during the stance phase and the quadriceps femoris contracts more strongly and during most of the stance phase as compared with walking on low heels. There are no obvious changes in the activity of the hamstrings, flexors of the hip, gluteus medius and maximus and erector spinae.

Author's address: Prof. J. JOSEPH, Department of Anatomy, Guy's Hospital, Medical School, *London* (England)

Medicine and Sport, vol. 6: Biomechanics II, pp. 192–196
(Karger, Basel 1971)

The Relation between Electrical Activity in Muscle and Speed of Walking and Running

M. MIYASHITA, H. MATSUI, and M. MIURA

Department of Physical Education, School of General Education, University of Nagoya, Nagoya

It has been reported that there is a close relation between energy cost and speed of walking and running [McDONALD, 1961; SARGENT, 1926]. As a matter of course, the amount of energy expenditure depends upon the quantity of muscular activity. The direct relation between muscular activity and speed, however, has never been revealed in the exercise of the whole body. The authors tried to find the relation between muscular activity and speed of walking and running from the view point of electromyography (EMG).

1. Method of the Integration of EMG Potentials

There are two main types of electrodes in the electromyographic study of muscle activity in the human. The one is the surface electrode and the other is the inserted electrode. Since each electrode has its characteristics, the surface electrode is generally applied to the study of physical movement. On the other hand, EMG (specially recorded by the surface electrode) can deal only with the qualitative analysis of muscle activity [BASMAJIAN, 1957]. Therefore, several investigators have tried to record the electrical activity of the muscle quantitatively. Particularly the integrated electrical activity has been recorded. LIPPOLD [1952] showed the mechanical integration with planimeter. BIGLAND and LIPPOLD [1954] performed the integration with a modified Miller's circuit.

A more complicated integrating circuit was made in this study. The integrator consisted of Miller's circuit. The response of the integrator to an input of constant voltage shows that the increasing wave form of output (integrator) is proportional to the magnitude of the input voltage such as 1, 2, and 5 mV [MATSUI *et al.*, 1969]. Using this integrator, the authors

[MIYASHITA *et al.*, 1969] conducted the experiments, the results of which indicated the same relation between muscular tension and integrated EMG under both isometric and isotonic conditions as INMAN *et al.* [1952] and BIGLAND and LIPPOLD [1954] already demonstrated. In fact, the relationship between the integrated electrical activity and the isometric tension is approximately proportional. In the case of isotonic contraction at constant tension during shortening, the average electrical activities increased with velocity. On the other hand, at constant velocity of shortening, the electrical activities increased according to the amount of tension.

2. The Relation between Integrated EMG and Speed of Walking and Running

Walking and running were performed on the motor-driven treadmill. Speeds of walking were 60, 80, 100, 120, 140 and 160 m/min and that of running were 150, 200, 250, 300, 350 and 400 m/min.

EMG was recorded using a pair of standard electrodes placed over the muscle in question. This was rectified and then led to the integrator. Rectus femoris and gastrocnemius were observed in the right leg. The movements of knee and ankle joint were recorded by electrogoniometer. Support and swing phases of the right leg were determined with the aid of semi-conductive strain gage placed in the shoe. The subject was a healthy adult man (height 180 cm and weight 72 kg) who was well accustomed to running on the treadmill.

Figure 1 shows the recording of EMG, integration, electrogoniogram and pressure of sole in walking at the various rates of speed. From those recordings it is clear that the walking process is repeated with every step. There is seen a peak of action potentials of rectus femoris and gastrocnemius in each cycle. At that time integration curves mount rapidly. Figure 2 represents the same recordings in running.

Integrated electrical activity per step was determined with the height of integration curve during one cycle. The value as measured above (rectus femoris plus gastrocnemius) was multiplied by the number of steps per minute and the product was the total electrical activities in walking or running for 1 min.

Figure 3 shows the relation between integrated electrical activity per minute (logarithm) and walking speed. MCDONALD [1961] calculated the experimental formula showing a relation between oxygen requirement (E′) and walking speed (V) with reference to the data reported by all researchers since 1927.

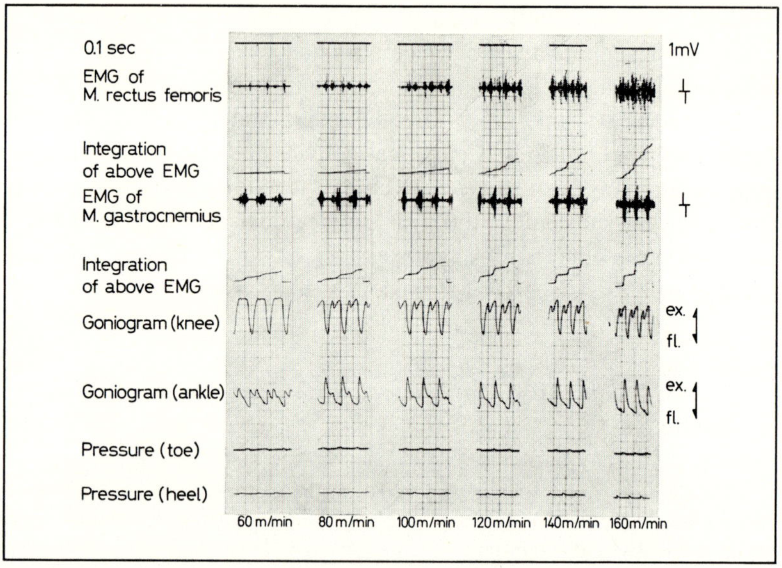

Fig. 1. EMG and its integration of rectus femoris and gastrocnemius with electrogoniogram and mechanical record of foot (walking at various rates).

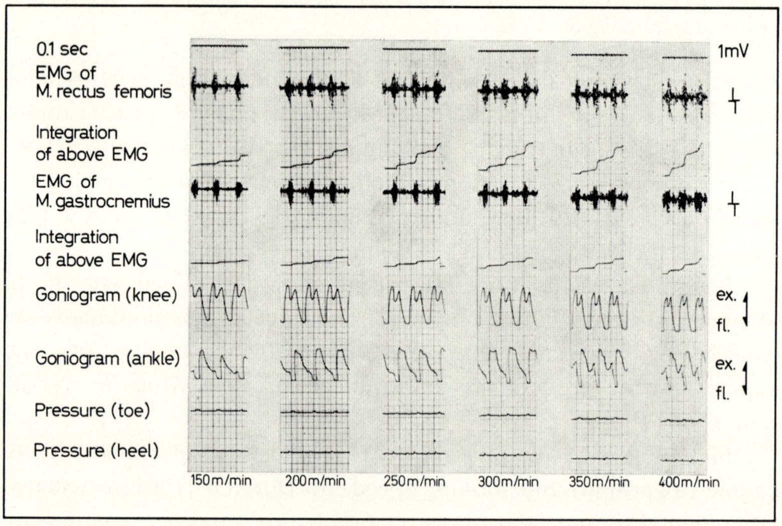

Fig. 2. EMG and its integration of rectus femoris and gastrocnemius with electrogoniogram and mechanical record of foot (running at various rates).

3, 4

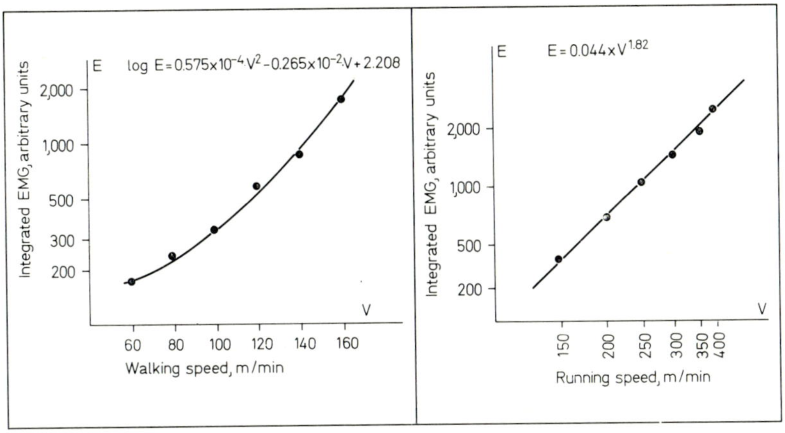

Fig. 3. The relation between integrated electrical activity (logarithm) and walking speed.

Fig. 4. The relation between integrated electrical activity and running speed (both in logarithm).

$$\log E' = aV^2 + bV + cW + d,$$

where W = body weight and a, b, c, d = constant. (1)

The result of this study was formulated as follows:

$$\log E = 0.575 \times 10^{-4} \cdot V^2 - 0.265 \times 10^{-2} \cdot V + 2.208, \quad (2)$$

where E = electrical activity.

Formula 1 was in accordance with formula 2.

Figure 4 shows the relation between integrated electrical activity per minute and running speed (both in logarithm). The relation was formulated as follows:

$$E = 0.044 \times V^{1.82} \quad (3)$$

Hitherto, SARGENT [1926] reported that the oxygen requirement per minute increased approximately as the 3.8th power of speed of running. The formula obtained in this study could not coincide with SARGENT's formula.

As reported above, there is a regular relation between electrical activity in muscles and speed of walking or running. That is, electrical activity increases according to the rate of speed. Comparing those results with the relation between energy cost and speed in walking or running, there is seen a good agreement in walking, but a little difference in running. The energy expenditure used in running increases more rapidly than muscular activity according to the rate of running speed. It can be, therefore, said that in the case of the moderate exercise such as walking, the energy cost increases in proportion to the number of muscle fibers participating in the exercise, but in the case of the strenuous exercise such as running, when the exercise needs the activity of the other functions in addition to the directly concerned muscles, the energy expenditure is necessarily higher than the amount of muscular activity according to the speed.

References

BASMAJIAN, J.V. and LATIF, A.: Integrated actions and functions of the chief flexors of the elbow. J. Bone Jt. Surg. *39:* 1106–1118 (1957).

BIGLAND, B. and LIPPOLD, O.C.J.: The relation between force, velocity and integrated electrical activity in human muscles. J. Physiol., Lond. *123:* 214–224 (1954).

INMAN, V.T. *et al.:* Relation of human electromyogram to muscular tension. Electroenceph. clin. Neurophysiol. *4:* 187–194 (1952).

LIPPOLD, O.C.J.: The relation between integrated action potentials in a human muscle and its isometric tension. J. Physiol., Lond. *117:* 492–449 (1952).

MATSUI, H. *et al.:* Electromyographic study on positive and negative works. I. Method of the integration of EMG potentials. Res. J. phys. Educ. *14:* 39–43 (1969).

MCDONALD, J.: Statistical studies of recorded energy expenditure of man. Nutr. abstr. Rev. *31:* 793–798 (1961).

MIYASHITA, M. *et al.:* Electromyographic study on positive and negative works. III. Relationship between force, velocity and integrated electrical activity in the positive and negative works. Res. J. phys. Educ. *14:* 98–102 (1969).

SARGENT, R.M.: The relation between oxygen requirement and speed in running. Proc. roy. Soc. B. *100:* 10–22 (1926).

Authors' address: Dr. MITSUMASA MIYASHITA, Dr. HIDEJI MATSUI, and Mr. MOCHYOSHI MIURA, Department of Physical Education, School of General Education, University of Nagoya, *Nagoya* (Japan)

Medicine and Sport, vol. 6: Biomechanics II, pp. 197–201
(Karger, Basel 1971)

Knee Extensor Static Strength Variations as a Function of Fibula-Femoral Angle

A. LINFORD and G. RARICK

Motor Performance Laboratory, University of Illinois, Champaign, Ill., and University of California at Berkeley, Berkeley, Calif.

The functional efficiency of the lower limb is a problem common to a broad spectra of academic areas; rehabilitation ergonomics, human factor engineering and physical education, to name but a few. More specifically, the measurement of static knee extensor strength has been frequently investigated and produced conflicting results as to the fibula-femoral angle at which maximum force is exerted. The most popular method of testing leg strength is one in which the subject adopts a standing position with knees bent and a belt is placed about his hips. He then attempts to extend his knees and his maximum extensor force is recorded.

In 1938 CARPENTER published what appears to be the earliest attempt to determine the knee angle producing maximum leg strength scores. Her results indicated that scores in the 115–124-degree range were optimal. It is unclear, however, if these angles were measured at the start of the test or at point of maximum strength recording. EVERTS and HATHAWAY [1938] and McCLOY and YOUNG [1954] also suggested testing angles of 120 and 130° respectively. HAXTON [1945] after a series of studies on cadavers, concluded that the leverage action of extensor muscles increases directly with joint extension. HUGH-JONES [1947] partially concurred with HAXTON in finding that for seated subjects knee extensor strength increased up to a point where the fibula-femoral angle was 1,965°. LINDEBURG [1964] found no significant difference in leg strength, in the inverted leg press position, with final knee angles between 120 and 160°. Inspection of his analysis of variance table, however, reveals an error in calculation. Recalculation of the *F* ratio, from reported data reveals it to be significant beyond the 0.001 level. BERGER [1966] in a study similar to the above found that maximum strength increased up to 140°. No significant differences were detected in the 80–130-degree range for subjects tested in the sitting position [CAMPNEY and WEHR, 1965]. The purpose of this study was to determine the extent to which static knee extensor strength, as conventionally measured by a leg and back dynamometer, varied as a function of fibula-femoral angle.

Method

Subjects. Twenty white male graduate students were used in this study. They were randomly selected from a university residence hall population.

Apparatus. The leg dynamometer used in this study was conventional, in so far as it consisted of a wide canvas belt, which was placed around the subject's hips. This belt was then inserted directly into a thrust bar which continued into the base platform. Thus, no bar was placed across the subject's thighs. Into the thrust bar were inserted four strain gauges mounted into a Wheatstone bridge circuit. This circuit incorporated a Honeywell 1508 Visocorder oscillograph and a Honeywell 133-2cc carrier amplifier. As the subject exerted force on the thrust bar a light beam was deflected on the light sensitive paper of the Visocorder. This produced a tracing of the force pattern exerted by the subject which could be measured by comparing it to a 500-lb. standard calibration deflection (fig. 1).

Independent variable. The following knee angle ranges were investigated: 115–124, 125–134, 135–144, 145–154 and 155–164°. These angles were measured with a goniometer at moment of maximum contraction. The knee joint angle was that formed by a line extending from the greater trochanter of femur to the lateral

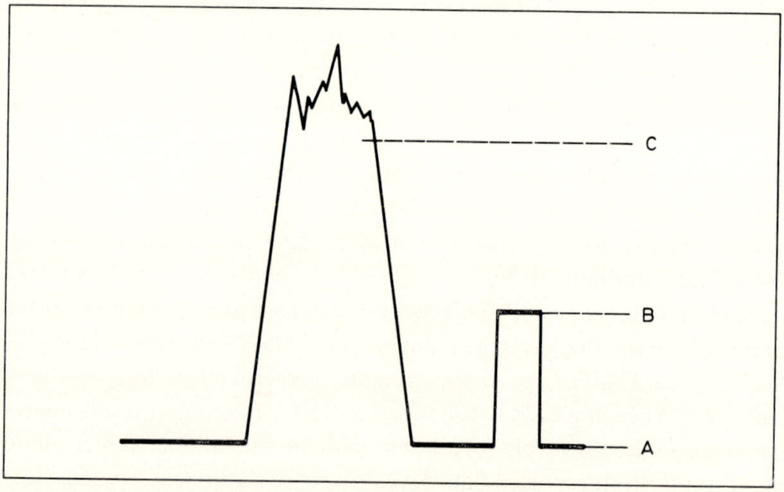

Fig. 1. Sample tracing of force curve in the measurement of leg strength. A = base line, zero force; B = calibration deflection (500 lb); C = force curve.

epicondyle of femur, and another line from the lateral malleolus through the head of the fibula. These points were determined by palpation.

Dependent variable. Static leg strength was measured in pounds. Reliabilities of dependent measures were 0.83 between trials and 0.93 between day means of two trials.

Testing. Ss were tested at one knee angle per day for a period of 5 days. Two trials were given per day and the mean of these two trials taken as the score for that knee angle. A Latin square design was used and subjects were randomly assigned to each cell of the Latin square. All subjects were tested at all knee angles. The same testers and standardized motivational instructions were used throughout and subjects were unaware of the hypothesis of the study. One day prior to the commencement of testing Ss were given a trial at every angle to familarize them with the test situation.

Results

The data were analyzed using analysis of variance for a type-2 mixed-model to determine strength differences between knee angles, order of testing, days and also the day and treatment interaction. These results (table I) revealed a significant difference between knee angles ($p < 0.001$) but no significant differences attributable to the other factors ($p > 0.05$). In order to identify the knee angles responsible for the significant treatment effect a Scheffé test for multiple *a potiori* comparisons was applied to all pairs of means. A decision rule, to reject the null hypothesis when differences beyond the 0.10 level were noted, was adopted to compensate for the conservative nature of this test. Table II lists the means for all angles and the differences between these means.

Table I. Analysis of variance of force readings taken at five knee angles

Source of variance	Sum of squares	df	Mean square	*F*
Treatments	2,139,168	4	534,792	34.48
Days	85,650	4	21,412	1.30
Days × treatments (w)	193,127	12	16,094	0.98
Error (w)	987,785	60	16,463	
Within-subjects	2,405,730	80		
Days × treatments (b)	365,305	4	91,326	0.823
Error (b)	1,665,284	15	111,019	
Between-subjects	2,030,589	19		
Total	4,436,319	99		

Table II. Significance of the differences in means of force readings obtained at different knee angles as shown by the Scheffé test

Angles (degrees)		Means (lb.)		Difference (lb.)
a	b	X_a	X_b	$X_a - X_b$
160	150	1,011.25	1,021.25	-10
160	140	1,011.25	1,016.25	-5
160	130	1,011.25	904.50	106.75[a]
160	120	1,011.25	639.50	371.75[a]
150	140	1,021.25	1,016.25	5
150	130	1,021.25	904.50	116.75[a]
150	120	1,021.25	639.50	381.75[a]
140	130	1,016.25	904.50	111.75[a]
140	120	1,016.25	639.50	376.75[a]
130	120	904.50	639.50	265.00[a]

[a] Significant at or beyond the 0.10 level. Confidence interval = 36.55, df 4, 60.

It can be seen that knee angles in the 135–164-degree range did not result in strength scores which were significantly different from each other. Scores within this range were, however, significantly superior to those obtained in any of the other fibula-femoral angle ranges. Angles in the 125–134-degree range produced significantly greater recorded strength than that obtained from the 115–124-degree range. Observation of the means would suggest that, in order to obtain maximum recorded leg strength score when the conventional method of testing is used, fibula-femoral knee angle of the subjects should be approximately 140–150° at moment of maximum effort.

Discussion

There are similarities between the results of this study and those of BERGER [1966] and LINDEBURG [1964], both of whom used an inverted leg press position. However, direct comparison is difficult because of the possible kinesiological difference in the action of the gluteus maximus in such a testing position. Kinesiological differences in testing position could also explain the disagreement between these and the results of both CAMPNEY and WEHR [1965] and HUGH-JONES [1947]. Comparison with CARPENTER'S [1938] work is impossible because of the fact that she did not specify if the angles she used were measured prior to pulling or at moment of maximum contraction. While HAXTON'S [1945] study was thorough, it is quite possible that variables which affect performance in the living subject, are not obvious in the cadavers.

The results of this study can be summarized as follows: when the conventional belt method of testing leg strength is employed: (a) Maximum force is recorded with fibula-femoral angles in the 135–164° range. Variations in angles within this range do not produce significant force differences. (b) As the angle increases from 115–134° recorded strength increases directly. The relationship between maximum score and other physical parameters (e.g., somatotype, fibula-femoral length ratios, etc.) may be worthy of further investigation.

References

1. Berger, R.A.: Leg extension force at three different angles. Res. Quart. amer. Ass. Hlth phys. Educ. *37:* 560–562 (1966).
2. Campney, H.K. and Wehr, R.W.: An interpretation of the strength differences associated with varying angles of pull. Res. Quart. amer. Ass. Hlth phys. Educ. *36:* 403–412 (1965).
3. Carpenter, A.: A study of leg angles in the measurement of leg lift. Res. Quart. amer. Ass. Hlth phys. Educ. *9:* 70–72 (1938).
4. Everts, E.W. and Hathaway, G.J.: The use of a belt to measure leg strength improves the administration of physical fitness tests. Res. Quart. amer. Ass. Hlth phys. Educ. *9:* 62–69 (1938).
5. Haxton, H.A.: A comparison of the extension of the knee and elbow joints in man. Anat. Rec. *93*: 279–286 (1945).
6. Hugh-Jones, P.H.: The effect of limb position in the seated subject on ability to utilize the maximum contractile force of the limb muscles. J. physiol., Lond. *105:* 332–344 (1947).
7. Lindeburg, F.A.: Leg angle and muscular efficiency in the inverted leg press. Res. Quart. amer. Ass. Hlth phys. Educ. *35:* 179–183 (1964).
8. McCloy, C.H. and Young, N.D.: Tests and measurements in health and physical education (Appleton-Century-Crofts, New York 1954).

Authors' addresses: Dr. Anthony G. Linford, Motor Performance Laboratory, University of Illinois, Room 97, *Champaign, IL 61820;* Dr. G. Rarick, University of California at Berkeley, *Berkeley, CA* (USA)

Medicine and Sport, vol. 6: Biomechanics II, pp. 202–207
(Karger, Basel 1971)

Analysis of Knee Function in Ascending and Descending Stairs

N. SHINNO

Department of Orthopedic Surgery, Tokushima University, Tokushima

The Relation between the Flexion Angle of the Knee and the Slope of the Stairs

In architecture, stairs are classified according to their slope as shown in table I.

Table I. Classification of the slope

Slope	Classification
0–15°	Sloping road
20–50°	Stairs (convenient)
50–75°	Special stairs
75–90°	Ladder

Of these stairs, those with a slope of 30–35° are the most convenient. To analyze to what degree the knee joint must be flexed in going up and down ordinary stairs (slope 20–50°), the theoretical relation between the maximum flexion angle of the knee joint and the slope of the stairs was calculated.

In figure 1, the slope of the stairs is φ, their width T, the height of each stair R, the flexion angle of the knee Θ, the angle of inclination of the body α, and the length of the lower limb 2a. If we assume that the knee joint is situated in the middle of the leg, we may write

$$\Theta = 2 \sin^{-1} \left\{ \cos \left(\sin^{-1} \frac{\sqrt{R^2+T^2} \cos \varphi}{2a} - \frac{\sqrt{R^2+T^2} \sin \varphi}{2a} \right\} . \quad (1)$$

$$(20° < \varphi < 50°)$$

The relation between the maximum flexion angle of the knee and the slope of the stairs was obtained using expression (1).

From the optimal height of stair measured by the consumption of oxygen, the expression $\sqrt{R^2 + T^2} = 33$ cm was obtained. Under conditions where energy consumption was minimal, the value $T-R = 12$ cm was obtained. T and R were calculated from the expression. The most convenient stair has a height of 16.6 cm, a width of 28.6 cm and a slope of about 30°. These results depend on the dimensions of the subjects.

Table II. Maximum flexion angle of the knee on the stairs

Slope of stairs	Maximum flexion angle of the knee	
20°	104°	
25°	99° 06	
30°	91°	optimum
35°	94° 42	optimum
40°	88° 40	
45°	84° 46	
50°	82° 12	

Substituting the expression $\sqrt{R^2+T^2}=33$ cm in expression (1), and assuming that the length of the leg is 82 cm (average for normal Japanese adults), the relation between φ and Θ is shown in table II. This theoretical analysis shows that to ascend or descend convenient stairs (slope 30°), the knee joint must be flexed 95°. By cinematography, it is generally recognized that in walking the maximum flexion angle of the knee is 140° and in running it is 120°. During ascending and descending stairs, it was calculated theoretically that the knee joint must flex about 90°. Knowledge of the flexion angle of the knee joint in normal movements of daily life is necessary for treatment of stiffened knee joints.

Analysis of Muscle Strength in Ascending and Descending Stairs

In ascending and descending stairs the knee joint is extended and flexed while supporting the body weight on one leg. The extent of the movement of the knee is from full extension to 90-degree flexion. In the semi-flexed position of the knee, the ligaments become loose, and, therefore, the stability of the knee depends mostly upon the strength of the quadriceps muscle acting through the patella. From the results and analysis of the muscle strength of the extensors and flexors of the knee in the stooping movement, the extensors, especially the medial and lateral vastus muscles, play the major role, while the flexors act as stabilizers. Experiments were made on how these muscles act in ascending and descending stairs.

Electromyography was used with surface electrodes for the rectus femoris, vastus lateralis, vastus medialis and biceps femoris. Experiments were made during ascending and descending stairs with a height of 20 cm, width of 25 cm and slope of about 40°. Ten normal Japanese males were tested. Electromyographic activity was recorded at a speed of the recording paper of 50 cm/sec.

The results obtained were as follows: in ascending, as soon as the body is supported by one leg and the other leg leaves the stair, the action of the quadriceps muscle increases. The muscular activity of the lateral vastus muscle is especially great. As in the stooping movement, the amplitude of the action potential of the biceps femoris was smaller than that of the extensors, and it is considered as a tonic NMU. In descending, the activity of the extensors of the knee were smaller than those in ascending, but greater than in the stooping movement, in which the quadriceps muscles of both legs support the body. Moreover, during descending, the amplitude of the biceps femoris was slightly greater than during ascending.

These results indicate that the extension and the flexion of the knee are performed while the equilibrium between two opposing forces on the patella is maintained, namely the body weight and the force of the quadriceps muscle while the body weight is maintained on one leg. Thus, if the quadriceps muscle becomes weak, the knee joint might become unstable when going up and down stairs, especially down.

Mechanism of Stabilizing the Knee in Ascending and Descending Stairs

To go up and down convenient stairs, the knee must be flexed from full extension to about 90° while supporting the body on one leg. Accordingly, in going up and down stairs, the stability of the knee must be greater than in walking or running. From the results obtained on the movement of the knee, it is considered that the stability of the knee is maintained by a combination of the actions of structural and functional stabilizers. In this section these stabilizers will be discussed.

1. Structural stabilizers of the knee. In the bony structure of the knee joint, the lower end of the femur is divided into complicated medial and lateral condyles, which form convex surfaces, while the articular surfaces of the tibial condyles opposite the femoral condyles are nearly flat. Results

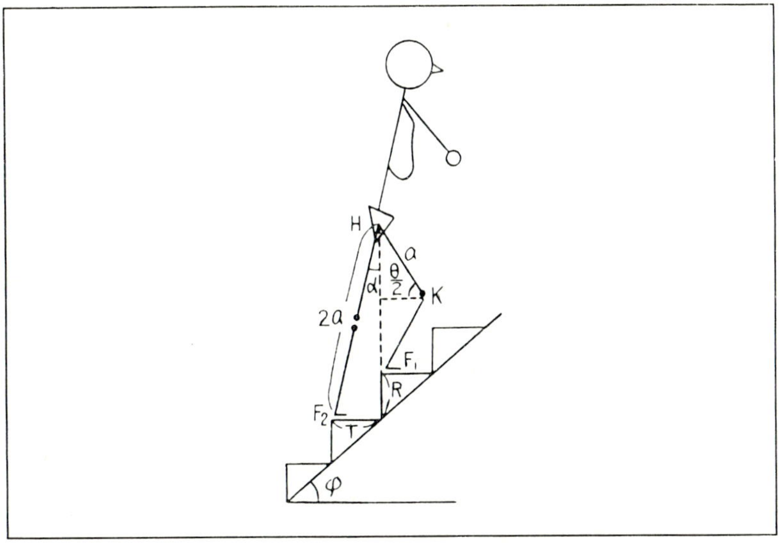

Fig. 1. See text.

obtained by slit-ray photography showed that the area of contact was greatest when the knee was fully extended, but that there was very little contact in the semi-flexed position although part of the body weight is supported by the condylar articular surfaces, most of the weight being supported by the patella and soft tissues.

The menisci which are interposed between the femoral and tibial condyles make up for the lack of contact between the condyles. They also distribute the load more equally over the condylar surfaces and increase the stability of the knee by acting as brakeshoes.

Studies by X-ray and slit-ray photography showed that the medial femoral condyle is larger than the lateral one. The femoral condyles cannot move straight forwards and backwards on the tibial condylar surfaces like two wheels. Therefore, during extension and flexion of the knee a rotatory, i.e. 'screwing-home' movement occurs.

In ascending stairs, the knee moves towards the 'screw-home' and the joint capsule, ligaments and other soft tissues become taut. On the other hand, in descending the knee moves from an extended to a semiflexed position, and the 'screw-home' loosens. The 'screw-home' mechanism also seems to contribute to the stability of the knee.

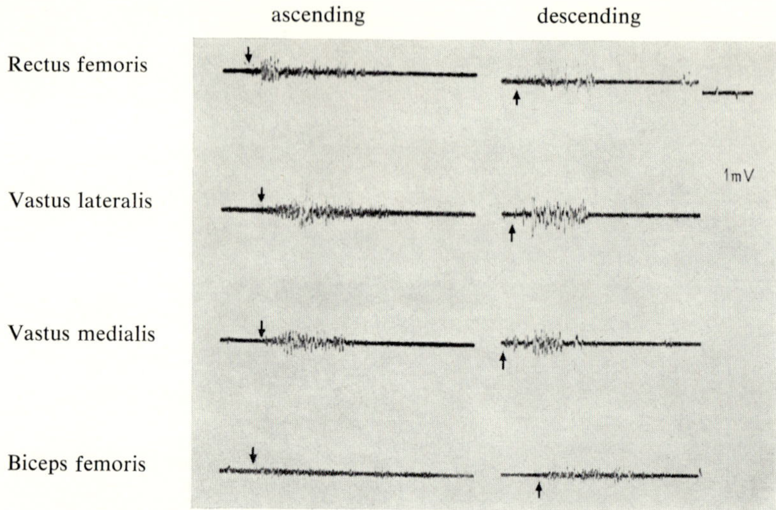

Fig. 2. Electromyogram in ascending and descending stairs.

2. *Functional stabilizers of the knee.* The functional stabilizers in the knee joint are the patella and the quadriceps muscle acting through it. The dynamic relation between the femoral and tibial condylar articular surfaces, the pressure of the femoral condyles on the tibial condyles becomes maximal when the knee is fully extended, and zero when the knee is flexed at an angle of 90°. When the knee is flexed more than 90°, the force tends to separate the two condylar surfaces.

The oppressing power of the quadriceps muscle on the patella opposes the oppressing power of the body weight on the patella, and thus the quadriceps stabilizes the knee joint from the front. By the relations between the oppressing power by the body weight and by the quadriceps on the patella, the knee joint is moved while supporting the body and these two oppressing powers maintain an equilibrium. The powerful quadriceps muscle acts effectively only through the patella.

Conclusion

The action of the knee in going up and down stairs was analyzed, and the results obtained were as follows:

(1) Theoretically, in going up and down convenient stairs, the knee joint must be flexed about 90°.

(2) In going up stairs, the action of the quadriceps muscle was recorded to be higher than during going down stairs. The amplitude of the action potentials from the flexors of the knee were small both during going up and down stairs.

(3) The knee joint is more unstable during going down than during going up stairs for the following reasons: When descending stairs, (a) the 'screwing-home' becomes loose, (b) the quadriceps muscle is stretched passively, (c) speed is increased by gravity.

Acknowledgement

The author expresses his sincere thanks to Prof. K. Yamada, of Tokushima University, for his guidance throughout this work.

References

Bowen, W. P. and Stone, H. A.: Applied anatomy and kinesiology, pp. 248–260 (Lea & Febiger, Philadelphia 1957).

Duvall, E. N.: Kinesiology, the anatomy of motion, p. 86 (Prentice-Hall, Philadelphia 1959).

Kitamuro, Y. and Shinno, N.: Functional significances of the patella in movement of the knee (in Japanese). J. jap. orthop. Ass. *9:* 1047 (1959).

Lewin, P.: The knee and related structures, pp. 28–59 (Lea & Febiger, Philadelphia 1952).

Design of the Building, High School Textbook (in Japanese), pp. 78–81 (Dept. of Education, Japan, 1956).

Shinno, N.: Functional significances of the quadriceps muscle in movement of the knee (in Japanese). Centr. jap. J. orthop. traumat. Surg. *3:* 116 (1960).

Shinno, N.: Statico-dynamic analysis of movement of the knee. I. Modus of movement of the knee. Tokushima J. exp. Med. *8:* 101 (1961). – II. Functional significances of the patella in movement of the knee. Tokushima J. exp. Med. *8:* 111 (1961). – III. Functional significances of the flexors and extensors of the knee. Tokushima J. exp. Med. *8:* 124 (1961). – IV. Functional significances of the menisci in movement of the knee. Tokushima J. exp. Med. *8:* 189 (1961).

Wells, K. F.: Kinesiology, pp. 232–246 (Saunders, Philadelphia/London 1960).

Author's address: Dr. N. Shinno, Department of Orthopedic Surgery, Tokushima University, *Tokushima* (Japan)

Medicine and Sport, vol. 6: Biomechanics II, pp. 208–211
(Karger, Basel 1971)

Critical Appraisal of Jumping as a Measure of Human Power

G. T. ADAMSON and R. J. WHITNEY

University of Leeds, Yorkshire, and Medical Research Council, Hampstead, London

Almost every edition of Res. Quart. amer. Ass. Hlth phys. Educ. contains studies or references to the *Sargent Jump* and without exception the term *power* is used concomitantly. For nearly 50 years the Sargent jump has been the most commonly used test of muscle power by physical educationists, coaches and athletes alike. SARGENT [1924] refers to 'the work done by causing the body to rise above the ground is the excess work done over that required merely to raise it from the squatting to the standing position, and this excess work consists of building up velocity, an accomplishment possible only when the rate of doing work (power) is above a certain minimum'. MCCLOY [1932] states that the Sargent jump when combined with an appropriate formula containing factors of age and weight predicts the *power* type of athletic ability, and that it is probably the best single measure we have of predicting explosive energy. Correlations are usually computed in research studies between the height jumped, and other variables usually associated with so-called *power* performance, e.g. strength or speed, and to date the explanation of the mechanics of the vertical jump has not been made clear. It would appear to be desirable to make clear what is implied by the word *power.*

The Concept of Muscle Power

Strictly, *power* is a carefully defined variable of applied mechanics, specifically designed for assessing the working capacity of an engine or motor. As used in mechanical engineering there is no necessity to draw a distinction between *power production* by the engine, and the *power dissipation* into an external load or resistance. In practical engineering application, production and dissipation can be regarded as simultaneous, and the power dissipation, which is readily measured by some form of brake dynamometer, can be taken as equal to the power production of the engine which it is desired to specify. Muscular activity which is performed at a sustained level such as pedalling on a bicycle ergometer can be similarly rated on a horse power basis.

For impulsive actions as jumping an entirely different situation exists which throws considerable doubt on the utility of power as a measure of muscular capacity. Consider the motion of a shot fired from a gun pointing vertically. Power for projecting the shot is derived from the expanding propellant charge, but it will be clear in this case that *power production is substantially completed before there is any appreciable dissipation of the power.*

The work done on the shot during its travel up the barrel is stored as kinetic energy of the shot $= \frac{1}{2} mv_0^2$. It would be possible to say that this work is done during the time, t_0, the shot takes to travel up the barrel, and that the average power of the expanding charge is

$$\frac{\frac{1}{2} mv_0^2}{t_0},$$

but it is not particularly useful to do this if we require to assess the efficiency or capacity of the system. This is more readily done by one of the following:

(a) Measuring the velocity, v_0, at which the shot leaves.

(b) Measuring the height to which the shot goes in doing work against gravity. This is a simple function of v

$$h = \frac{v_0^2}{2g}$$

(ignoring air resistance).

(c) Measuring the reaction (F) of the barrel on the ground–the area under the $F \times t$ curve will be equal to the momentum (mv_0) imparted to the shot.

Before leaving this example it is necessary to note the following points, which will be relevant to the jumping situation:

(i) If, after some time t_1 after firing the shot is still in the barrel and the instantaneous velocity of the shot is v_1 and the reaction on the ground is F_1, the power production up to that time is clearly

$$\frac{\frac{1}{2} mv_1^2}{t_1}$$

and *not* $F_1 v_1$.

(ii) The power dissipation in working against gravity after the shot leaves the gun has no particular relevance to the action of firing or the as-

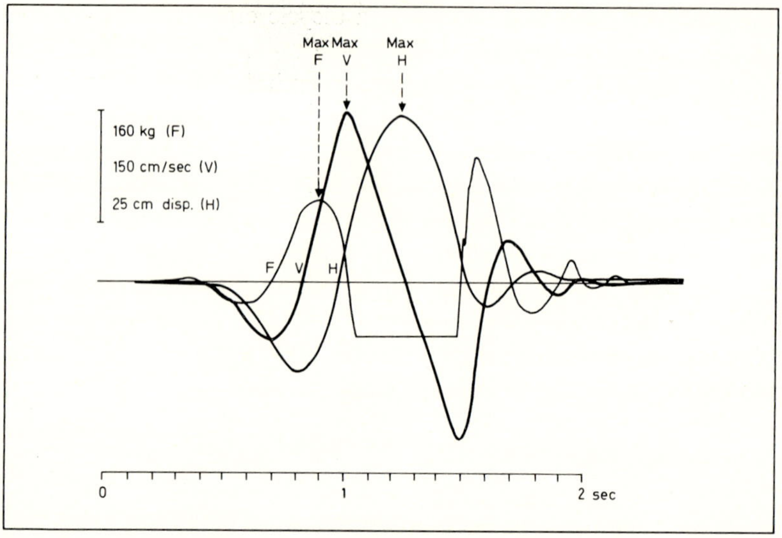

Fig. 1. Smoothed force/time, velocity/time, distance/time record of a vertical jump performed on the force platform at the Human Biomechanics Laboratory, Medical Research Council, Hampstead, London.

sessment of the charge. If at any time the upward velocity is v, then the power dissipation is mgv.

(iii) The power is not really dissipated in the sense that engine or bicycle ergometer power is dissipated (as heat in the brake); it is stored as potential energy of the shot and, ideally, the whole kinetic energy of $\frac{1}{2}mv_0^2$ is regained when the shot falls to the ground.

The whole conclusion of this dissertation must be that power, in the strict mechanical sense, will not be a useful parameter of impulsive muscle action. The real measure of the action is the $F \times t$ impulse itself and the manner in which co-ordinated muscle action can generate the largest possible impulse.

Reference to the vertical jump force/time trace on which are superimposed the simultaneous velocity/time and distance/time graphs obtained by successive integration will serve to make clear some general points which are relevant to the techniques of recording and analysis.

It will be noted that peak force precedes peak velocity, as would be expected on purely mechanical considerations. DAVIES and RENNIE [1968] have regarded the product of instantaneous force and velocity as indicating the instantaneous power production. There is the almost intuitive objection that the instantaneous velocity of movement of the centre of mass is not due to the current force but to the preceding force time integral. The integral itself is the *impulse* generated by the jumping action, the size of the impulse is directly related to the subsequent success of the jump, and the shape of the impulse is likely to give the best indication of the muscular activity associated with its generation. The area of the positive pulse must be as large as possible, and there are three theoretical possibilities which are exploited to different extents by different jumpers: (i) increase of peak force, (ii) increase of pulse duration, (iii) squaring of the pulse-particularly by making the rising phase as steep as possible by relating the pulse slope (dF/dt) to the actual observed body motion (film).

Consideration of the means by which the above are achieved is not the purpose of this article.

Conclusions

The essential conclusion from this argument is not only that the use of *'power'* or *'power'/weight* in connection with impulsive actions is unjustified, but that power is not a useful concept for such actions.

References

DAVIES, C.T.M. and RENNIE, R.: Human power output. Nature, Lond. *217:* 770–771 (1968).

MCCLOY, C.H.: Recent studies on the Sargent jump. Res. Quart. amer. Ass. Hlth phys. Educ. *3:* 235–242 (1932).

SARGENT, L.W.: Some observations in the Sargent test of neuromuscular efficiency. Amer. phys. Educ. Rev. *29:* 47–56 (1924).

Authors' addresses: Mr. GRAHAM T. ADAMSON, Department of Physical Education, University of Leeds, *Yorkshire;* Dr. REGINALD J. WHITNEY, Human Biomechanical Laboratory, Medical Research Council, Holly Hill, Hampstead, *London N.W. 3* (England)

Medicine and Sport, vol. 6: Biomechanics II, pp. 212–218
(Karger, Basel 1971)

Self-Rotation of Animate Beings[1]

T. R. Kane

Stanford University, Stanford, Calif.

It is a well-known fact that both animals and humans can perform well-controlled self-rotation maneuvers while falling freely for short periods of time in the neighborhood of the surface of the earth. To determine the extent to which it may be possible to exploit this capability for the purpose of controlling the orientation of a human subject during prolonged periods of free fall, which is of interest in connection with manned space flight, dynamical analyses of a number of self-rotation maneuvers have been performed. Specifically, an analytical theory dealing with the righting movements of falling cats has been constructed, and the validity of this theory has been tested by reference to photographs; and pitch, yaw, and roll motions generated by arm and leg motions of humans have been studied analytically and, in the case of yaw motions, experimentally. Currently, experiments in a 'zero-g' flight of an aircraft and in an orbital space vehicle are being planned.

1. Righting Movements of Falling Cats

Falling cats frequently manage to land on their feet, even when released from rest while upside-down. In order to gain some insight into the method used by cats to accomplish this feat, the attempt was made to devise a kinetic 'explanation', i.e. to find a mechanical system whose motion, when proceeding in accordance with the laws of Newtonian mechanics, could mirror those of the animal. To test the validity of the explanation, drawings representing the mechanical system, produced by a computer programmed to solve the differential equations of motion, were superimposed on still-photographs of a falling cat.

The mechanical system selected for this purpose consisted of two rigid, axially symmetric bodies, A and B, with one common point, O, as shown in figure 1. A rather

1 This work was supported by NASA under NGR-05-020-209.

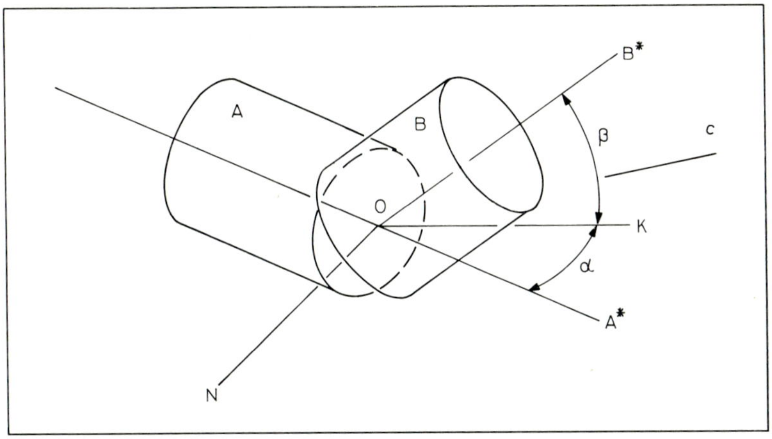

Fig. 1

complicated motion of B relative to A was postulated in order to accommodate two requirements imposed by physiological considerations, namely that neither twisting nor backward bending could take place to any considerable extent. This was accomplished as follows (fig. 1): It was postulated that B*, the symmetry axis of B, moves on the surface of a cone of semi-vertex angle β, the axis of the cone, line K, being fixed in body A at an angle β with the symmetry axis A* of A, and that the angular velocity of B relative to A satisfies the equation $\boldsymbol{\omega} \cdot \mathbf{C} = \mathbf{0}$, where $\mathbf{C}$ is a unit vector parallel to the bisector of the angle A* O B*. This condition eliminated undesirable twisting; and excessive backward bending was avoided by limiting the ratio of β to α.

Details of the dynamical analysis of this system may be found elsewhere [KANE and SCHER, 1969]. However, it may be worthwhile to comment further on the condition $\boldsymbol{\omega} \cdot \mathbf{C} = \mathbf{0}$, used to eliminate twisting, for this not only played a role in the investigation under consideration, but also in those concerned with motions of humans.

To discuss changes in the relative orientation of two rigid bodies A and B which have a common point O, it is convenient to fix mutually perpendicular unit vectors $\mathbf{a}_1$, $\mathbf{a}_2$, $\mathbf{a}_3$ in A and $\mathbf{b}_1$, $\mathbf{b}_2$, $\mathbf{b}_3$ in B; to define as the 'neutral' state of the system that state in which $\mathbf{a}_i = \mathbf{b}_i$, $i = 1, 2, 3$; and to designate as the 'neutral axis' of the system a line that passes through O and is parallel to $\mathbf{a}_1$ (and hence $\mathbf{b}_1$) when the system is in the neutral state. According to EULER's theorem on rotations about a point, it does not matter how a change in the orientation of B relative to A is actually produced: there always exists a line L passing through O such that the same change could be effected by simply rotating B about L. This fact makes it possible to give precise meanings to the terms 'bending' and 'twisting' in the present context. When the system is not in the neutral state, it is said to be 'purely bent' or 'untwisted' if L is perpendicular to both $\mathbf{a}_1$ and $\mathbf{b}_1$; 'purely twisted' or 'unbent' if L is parallel to both $\mathbf{a}_1$ and $\mathbf{b}_1$; and partially bent and twisted in all other cases.

Whenever the system is untwisted, its configuration can be described completely by the two angles Θ_1 and Θ_2 shown in figure 2; and the direction cosines relating the unit vectors fixed in the two bodies then have the values shown in table I, where C_i and S_i denote cos Θ_i and sin Θ_i, respec-

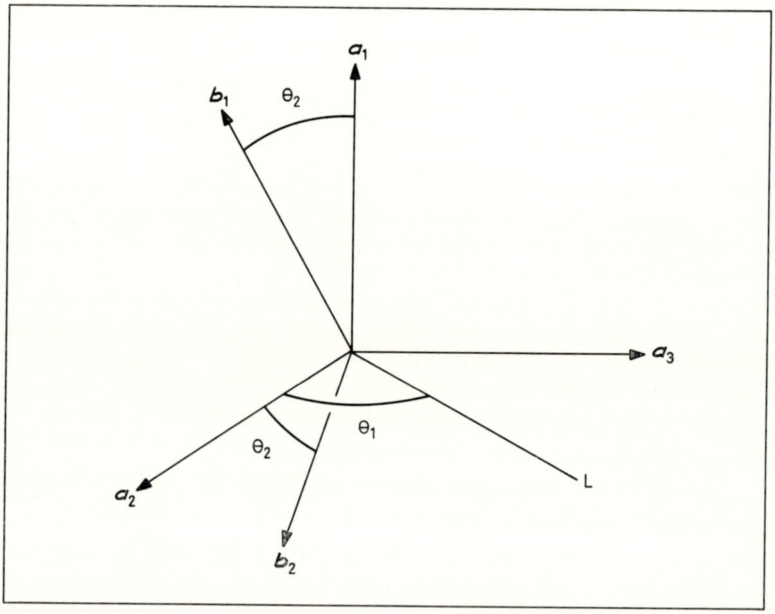

Fig. 2

Table I

	$\mathbf{b}_1$	$\mathbf{b}_2$	$\mathbf{b}_3$
$\mathbf{a}_1$	C_2	$-S_1S_2$	C_1S_2
$\mathbf{a}_2$	S_1S_2	$C_1^2+S_1^2C_2$	$S_1C_1\ (1-C_2)$
$\mathbf{a}_3$	$-C_1S_2$	$S_1C_1\ (1-C_2)$	$S_1^2+C_1^2C_2$

tively. The most general configuration change can be accomplished by aligning $\mathbf{b}_i$ with $\mathbf{a}_i$ for i=1,2,3 and then performing successive rotations of B relative to A of amounts Θ_1, Θ_2, and Θ_3 about axes parallel to $\mathbf{b}_1$, $\mathbf{b}_2$, and again $\mathbf{b}_1$, leading to the configuration indicated in figure 3. The associated direction cosines are tabulated in table II, and it is readily apparent that corresponding entries in table I and table II become identical if $\Theta_3 = -\Theta_1$. Hence, the system remains untwisted during any motion such that $\Theta_3 = -\Theta_1$. Now, the angular velocity $\boldsymbol{\omega}$ of B relative to A is given by

$$\boldsymbol{\omega} = (C_2\dot{\Theta}_1 + \dot{\Theta}_3)\mathbf{b}_1 + (S_2S_3\dot{\Theta}_1 + C_3\dot{\Theta}_2)\mathbf{b}_2 + (S_2C_3\dot{\Theta}_1 - S_3\dot{\Theta}_2)\mathbf{b}_3,$$

and, from table II,

$$\mathbf{a}_1 = C_2\mathbf{b}_1 + S_2S_3\mathbf{b}_2 + S_2C_3\mathbf{b}_3.$$

Table II

	$\mathbf{b}_1$	$\mathbf{b}_2$	$\mathbf{b}_3$
$\mathbf{a}_1$	C_2	S_2S_3	S_2C_3
$\mathbf{a}_2$	S_1S_2	$C_3C_1 - S_1C_2S_3$	$-S_3C_1 - S_1C_2S_3$
$\mathbf{a}_3$	$-C_1S_2$	$C_3S_1 + C_1C_2S_3$	$-S_3S_1 + C_1C_2C_3$

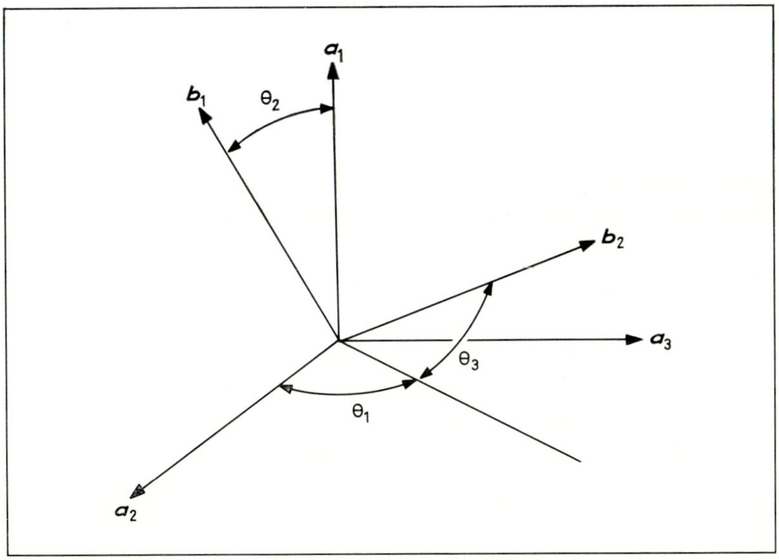

Fig. 3

With the help of these relationships one finds that

$$\boldsymbol{\omega} \cdot (\mathbf{a}_1+\mathbf{b}_1)=(\dot{\Theta}_1+\dot{\Theta}_3)\ (1+C_2),$$

and this shows that, whenever $\boldsymbol{\omega} \cdot (\mathbf{a}_1+\mathbf{b}_1)=0$, then $\dot{\Theta}_3= -\dot{\Theta}_1$, and hence $\Theta_3= -\Theta_1$, since Θ_1 and Θ_3 both vanish when the system is in the neutral state. Furthermore, $\mathbf{a}_1+\mathbf{b}_1$ is parallel to the bisector of the angle between $\mathbf{a}_1$ and $\mathbf{b}_1$. Consequently, to obtain relative motions during which the system remains untwisted, it is sufficient to make $\boldsymbol{\omega} \cdot \mathbf{C}=0$, where $\mathbf{C}$ is parallel to the bisector of the angle between $\mathbf{a}_1$ and $\mathbf{b}_1$.

2. *Human Motions*

A motion of a human being during which the torso rotates about an axis normal to the plane of symmetry of the torso is called 'pitching'. During free fall, pitching can be produced by holding the arms straight at the elbows and performing symmetrical rotational motions of the arms relative to the torso. If each arm is regarded as possessing an axis of symmetry, suitable rotational motions are generated by moving the symmetry axis of each arm on the surface of a cone whose axis passes through the shoulder joint; and if one requires that the motion of each arm relative to the torso be a pure bending motion in the sense discussed in the preceding section, then the effectiveness of the maneuver becomes solely dependent on the orientation of the cone axis relative to the torso and on the semi-vertex angle of the cone.

Numerical solutions of the differential equations of motion of a system moving in this manner reveal the following facts: (1) It is advantageous to let the cone axis lie in the plane determined by a spinal axis and a line joining the shoulder joints) at a downward angle of about 30° with the latter line. (2) The effectiveness of the maneuver increases with increasing cone semi-vertex angle. (3) Holding the legs in a 'tucked' position, rather then straight, has a significant effect. (4) Under optimum conditions, torso rotations of about 33° can be produced by means of a single rotation of the arms. (5) Holding weights in the hands can be very helpful. For example, with a 5-pound weight in each hand, and with the legs in a tucked position, a torso rotation of nearly 60° can result from one cycle of the maneuver.

Yaw motions, that is, rotations of the torso about an axis that is essentially parallel to the spine, can result either from arm or leg motions. A

2-phase maneuver that can be performed either with the arms or with the legs proceeds as follows when the legs are used: In the first phase, the legs are placed as when taking a step in ordinary walking, and the axis of each leg is then moved on the surface of a cone whose axis is parallel to the spine. In the second phase, the positions of the legs are interchanged, the axes of the legs moving in planes parallel to the plane of symmetry of the torso until the system has been restored to its initial configuration. The torso rotates in opposite directions during the two phases of this motion. However, the rotation produced during the first phase exceeds that resulting from the second phase, so that a net rotation is realized.

Numerical solutions of the relevant differential equations show that one cycle of this maneuver can produce a torso rotation of up to 70° when the legs are used, whereas only about 30° of rotation can be obtained by means of the corresponding arm motions. However, the arm motions lend themselves more readily to experimentation because an experiment involving these motions can easily be performed on a trampoline. In doing so, it is important to take precautions to eliminate 'initial condition' effects; i.e., if the subject knows that he must attempt to turn to the right after leaving the trampoline, he may unintentionally use the trampoline to put himself into a state of rotational motion to the right prior to the initiation of the maneuver. To avoid this in experiments performed at Stanford University (and recorded on motion picture film), the desired direction of rotation was indicated to the subject by means of a verbal or visual command after he had lost contact with the trampoline. In order to have enough time to perform even one complete maneuver, the trampolinist thus had to jump rather high, but experienced gymnasts were able to do this without difficulty.

Rotational motions of the torso about an axis lying in the plane of symmetry of the body and perpendicular to the spine, or 'roll' motions, cannot be produced so easily as can pitch and yaw motions. However, any roll reorientation can be attained, in a variety of ways, by means of successive pitch and yaw maneuvers. For example, a 90° roll to the left results from either of the following two sequences of 90° rotations: left yaw, forward pitch, and right yaw; or forward pitch, right yaw, and backward pitch. Of course, this process is rather time consuming. Hence, only an experiment performed during an orbital flight can be expected to reveal the extent to which it is feasible to execute such a sequence of maneuvers. On the other hand, the pitching motion discussed earlier, while unsuitable for a trampolinist, can probably be performed quite effectively during the approximately 30 sec of 'zero-g' flight that can be produced in an aircraft. Accordingly, preparations are currently being made for such a flight. Once it has been established in this way that both the pitch and the yaw maneuvers proceed essentially as predicted on theoretical grounds, orbital experiments will be required to study such physiological and psychological problems as fatigue, ability to select a suitable sequence of maneuvers, and optimum performance rate. Plans for such experiments are being made at the present time.

Reference

KANE, T.R. and SCHER, M.P.: A dynamical explanation of the falling cat phenomenon. Int. J. Solids Struct. *5:* 663–670 (1969).

Author's address: Dr. THOMAS R. KANE, Stanford University, *Stanford, CA 94305* (USA)

c) Human Motion in Sports

Medicine and Sport, vol. 6: Biomechanics II, pp. 220–224
(Karger, Basel 1971)

Effects of Altered Slope and Speed on the Biomechanics of Running[1]

R. C. NELSON and R. G. OSTERHOUDT

Biomechanics Laboratory, Pennsylvania State University, University Park, Pa.

Introduction

The biomechanics of running has attracted the attention of researchers for many decades. This interest has been stimulated by the fundamental nature of running and the applicability of the results of such research to a variety of popular sports. Although many aspects of the mechanics of running have been investigated to date, few studies have dealt directly with the changes in biomechanics which accompany an increase in running velocity and few, if any, studies have examined the changes associated with running on other than horizontal surfaces. The lack of research on hill running is unfortunate since cross country and marathon runners are required to perform on a variety of slopes. Consequently, this investigation represented an initial attempt to study the influence of running slope on the biomechanics of running.

Purpose

The purpose of this study was to quantify and compare the biomechanics of running at three speeds and on three slopes. Speeds selected for cinematographic analysis were 11, 16 and 21 ft/sec while the slopes were horizontal, minus 10% grade (downhill) and plus 10% grade (uphill). These speeds and slopes were chosen because they represented conditions present in cross country running. Although a total of 14 variables were studied only the following 4 were presented in this paper: stride length, stride rate, period of support and period of non-support.

1 This study was supported by the USOE Regional Research Program, Project No. 8-B-012.

Summary of Research Literature

Since the authors were unable to locate any studies concerned with uphill or downhill running, only studies dealing with the influence of altered speed are reviewed. A number of studies have shown that both stride length and stride rate increase as the running velocity increases. A representative sample of these studies included those by FENN [1929, 1930], CURETON [1935], DITTMER [1962], CAVAGNA *et al.* [1965], NEUMAN [1967a, b], COOPER and GLASSOW [1968]. One exception to these results was reported by ROMPOTTI [1956] who concluded that running velocity had no effect upon stride length.

Studies in which the periods of support and non-support were investigated have generally indicated that as the velocity increases the period of support becomes shorter while the period of non-support becomes longer. Investigators reporting these findings included DITTMER [1962], HOPPER [1964], NEUMAN [1967a, b] and COOPER and GLASSOW [1968].

Methodology

Sixteen experienced runners ranging in age from 18 to 22 were filmed while running indoors on a specially constructed platform which could be elevated to produce changes in slope. A pacer unit which provided a moving reference along the edge of the platform was employed to assist the runners in maintaining the desired running speeds. Two practice days were used to familiarize the subjects with the test conditions. Four test days were used, two for trials on the horizontal surface and two for both up and downhill trials. Three runs were made by each subject for each combination of speed and slope while only the third trial was filmed.

A Hy-Cam, 16-mm high-speed camera was used which contained internal timing lights used for calibration. Although the camera was set at 160 fps the calibration procedures revealed the actual speed to be 153 fps. A Vanguard Motion Analyzer was used for purposes of determining the quantitative data for calculation of stride length and rate and periods of support and non-support.

Results

The data were analyzed using 3-way analysis of variance procedures. When significant differences were noted for speed and/or slope the Tukey test was employed to evaluate the differences between each pair of conditions. The group means are presented graphically in figures 1-4. Since variability among the subjects was expected to be quite large, no attempt was made to compare the individuals any further.

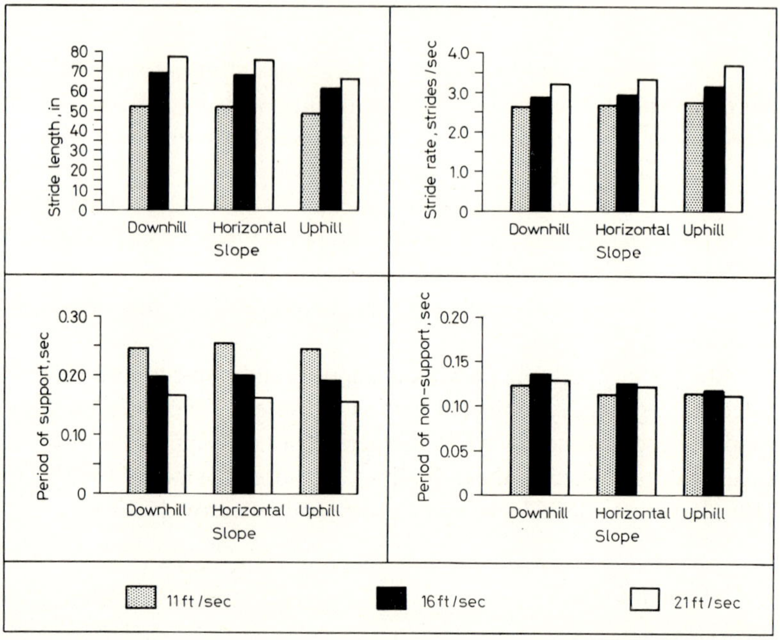

1, 2

3, 4

Fig. 1. Group means for stride length within slopes and across speeds. (Performed overground).

Fig. 2. Group means for stride rate within slopes and across speeds. (Performed overground).

Fig. 3. Group means for period of support within slopes and across speeds. (Performed overground).

Fig. 4. Group means for period of non-support within slopes and across speeds. (Performed overground).

The analysis of the stride length data revealed significant differences among individuals, speeds and slopes (fig. 1). Further evaluation revealed that all speeds as well as all slopes produced significant variations in stride length. As in previous studies increases in velocity were paralleled by lengthening of the running stride. However, the results indicated that downhill running produced the longest strides followed by horizontal running with the shortest strides produced during uphill running.

The stride rate results (fig. 2) revealed that significant increases occurred as the speed of running became faster. As would be expected, the slopes also produced changes in stride rate. The lowest stride rate occurred during downhill running while the highest rate was observed during uphill running. This could be anticipated in the light of the stride length results.

The data analysis for period of support also showed (fig. 3) significant differences among the means for slopes and speeds. Further evaluation revealed that all speeds were significantly different with the period of support decreasing as the speed increased. Comparison of the means for slopes revealed that the uphill condition produced significantly shorter periods of support than either horizontal or downhill but that the latter two slopes were not significantly different.

Significant differences for the period of non-support were also noted between all slopes and speeds (fig. 4). However, the performance over the three speeds did not reveal a consistent pattern as the longest period of non-support was observed at the middle velocity. This finding suggests that a non-linear relationship may exist between running speed and period of non-support. The influence of slope (fig. 4), however, was more clear since the time of non-support decreased consistently as the slope became more positive (from –10% to 0 to +10% grade).

In summary, increases in running velocity were associated with an increase in both stride length and stride rate and a decrease in the period of support. However, the period of non-support increased from 11 to 16′/sec but then decreased from 16 to 21′/sec. In general these findings are in agreement with most previous research dealing with the effects of changes in speed. Alteration of the running slope from a negative grade of 10% to a horizontal surface and then up to a positive 10% grade produces decreases in stride length and periods of support and non-support while causing an increase in stride rate. Since this study is one of the few concerned with the effects of altered slope it is recommended that future research be conducted in which a wider range of slopes and speeds are investigated.

Conclusion

Alterations in both running slope and speed produce significant, predictable changes in stride length, stride rate, period of support and period of non-support.

References

CAVAGNA, G. A.; MARGARIA, R., and ARCELLI, E.: A high-speed motion picture analysis of the work performed in sprint running. Ency. cinemat. *5:* 309–319 (1965).

COOPER, J. M. and GLASSOW, R. B.: Kinesiology; 2nd ed. (Mosby, St. Louis 1968).

CURETON, T. K.: Mechanics of track running. Schol. Coach. *4:* 7–10 (1935).

DITTMER, J.: A kinematic analysis of the development of the running pattern of grade school girls and certain factors which distinguish good from poor performance at the observed ages; unpublished Master's thesis, Madison (1962).

FENN, W. O.: Frictional and kinetic factors in work of sprint running. Amer. J. physiol. *12:* 583–611 (1929). – Work against gravity and work due to velocity changes in running. Amer. J. physiol. *93:* 433–462 (1930).

HOPPER, J.: The mechanics of arm action in running. Track tech. 520–522 (1964).

NEUMAN, J.: Casova charakteristika sprinterskeho behu 12–13 letych chlapcu (Chronometrical characteristics of the sprint of 12–13-year-old boys). Teor. praxe tel. vych. *15:* 292–297 (1967a). – Rychlost behu v prvnich metrech sprinterske trate (The speed in running the first meters of track sprinting). Teor. praxe tel. vych. *15:* 464–471 (1967b).

ROMPOTTI, K.: A study of stride length in running. International track and field digest, 249–256 (1956).

Authors' address: Dr. RICHARD C. NELSON and Mr. ROBERT G. OSTERHOUDT, Biomechanics Laboratory, *University Park, PA 16802* (USA)

Medicine and Sport, vol. 6: Biomechanics II, pp. 225–231
(Karger, Basel 1971)

The Mechanics of the Sprint Start

A. H. Payne and F. B. Blader

University of Birmingham, Birmingham

Equipment

The force plates [Blader and Payne 1967; Blader, 1968] which form the starting blocks are supported on semi-circular sensing elements fitted with wire resistance strain gauges. These are connected into Wheatstone Bridge circuits and the unbalanced signals are amplified and fed to an ultra-violet (UV) recorder giving as output a series of lines on moving photographic paper. The traces from each foot represent the force normal to the block surface, the force tangential to the block surface and the torque or couple about a horizontal axis across the block (fig. 1). Connected to other galvanometers in the UV recorder are a microphone to pick up the sound of the starter's gun, three photo-electric timing beams to record the athlete's progress down the track and a large visual clock to synchronise ciné film and paper recording.

Method

At least 6 starts by each of 17 British athletes were recorded, the same starter acting throughout. Competition conditions were approximated by having other athletes starting in nearby lanes at the same time. The analyses of the forces were carried out by semi-automatic transference to paper tape suitable for subsequent digital computer processing. The magnitudes, angles and points of application of the resultant forces for each foot and for both feet at any instant and the mean horizontal acceleration while the athlete was driving on the blocks were calculated (e.g. fig. 2).

The location of the centre of gravity of the athlete at various instantaneous positions during the drive from the blocks was determined using a centre of gravity board supported on 4 weighing scales, and the force vector corresponding to each of these positions was drawn to scale on a print of the appropriate ciné frame. The line of the resultant force relative to the position of the centre of gravity was then evident (fig. 3).

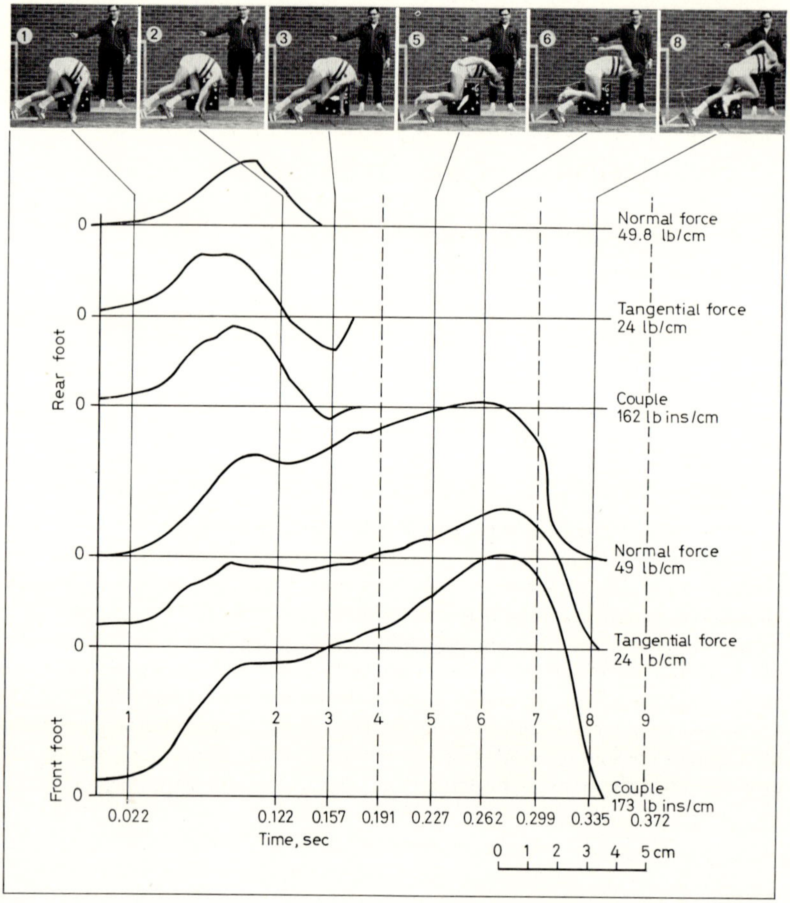

Fig. 1. Replot of an actual trace of a single sprint start showing force patterns in time with selected frames from the ciné film to illustrate body posture. For each foot is shown three force traces: (i) the force at right angles to the block face; (ii) the force along the block face, with downwards forces showing positive; and (iii) a couple about a transverse horizontal axis where a clockwise force as seen in the photographs shows positive.

Time measurements were obtained from the UV recorder traces since the instrument is capable of superimposing on the trace vertical timing lines accurate to 0.0001 sec. Taking into account HENRY's [1952] arguments on the timing of sprint starts the autors used photo-electric timers at 20 ft (6.10 m), 30 ft (9.14 m) and occasionally at 50 ft (15.24 m). These were suitable for normal starts with athletes using their accustomed block spacings, but more timers further down the track may be necessary when unusual block conditions are tested.

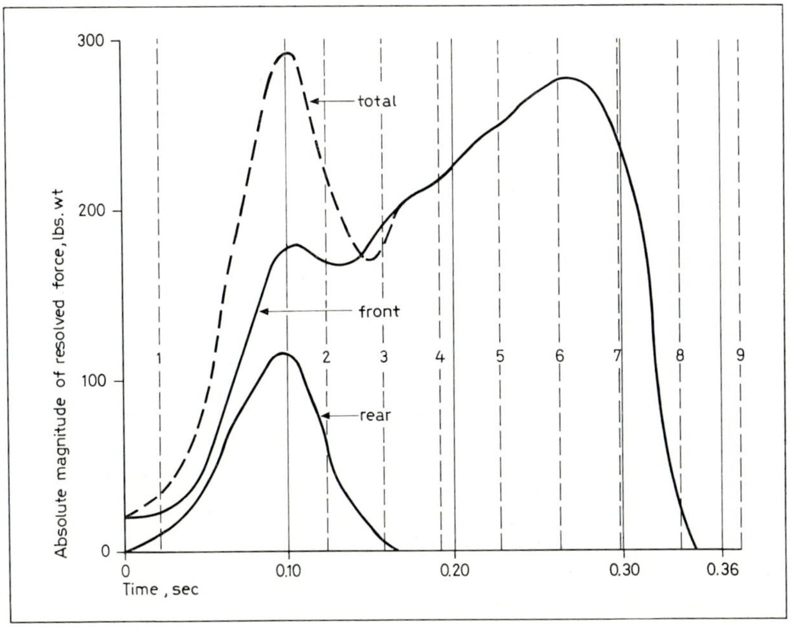

Fig. 2. Calculated values for the magnitude of the resolved forces for front foot, rear foot and both feet. These values were computed from the trace in figure 1.

Results and Discussion

The Patterns of the Force Traces

In general both rear and front feet started to exert forces upon the blocks at the same instant, rarely being separated in time by more than 0.01 sec. The heights of the rear foot peaks varied a great deal from athlete to athlete, but those with the best starts usually had strong rear leg action. However, as impulse is more meaningful than peak force, it was the front foot with its much longer contact time which provided the greater part of the acceleration of the body. The shape of the rear foot trace was usually symmetrical though there were one of two exceptions, while that of the front foot tended to show two peaks with the second higher than the first (fig. 1 and 2).

Fig. 3. This sequence from the ciné film shows the relationship between the evoked force vectors (arrow), drawn to scale, and the athlete's centre of gravity (×).

Resultant Force Evoked and the Centre of Gravity

It is apparent in figure 3 that the force evoked by this sprinter acts initially in front of his centre gravity, tending to rotate his body up to a vertical position (anticlockwise in fig. 3), but later in the start, during the time that the free leg is reaching forward in the first step, the force acts behind the centre of gravity tending to rotate the athlete downwards (clockwise). Presumably the rotation upwards is predominant in the period on the blocks as it must be in the next several steps if the athlete is to gradually assume a more upright position for the rest of the race.

Time Measurements

An average time of about 0.09 sec has been obtained between the sound of the gun and the first rise shown by the force trace. This is considerably faster than reaction times of the same subjects obtained by conventional methods and it is probable that the authors were here measuring not total reaction time but the so-called 'pre-motor' periods of total reaction time. This theory was supported by the fact that the first rise of the force trace did not coincide with a perceptible movement by the athlete – indeed many false starts were not apparent until the traces were studied. Further tests are being carried out at present using variable 'hold' times between the command 'set' and the firing of the gun in order to eliminate the possibility of the athlete anticipating the gun.

Other time measurements ranged as shown in table I for the different athletes tested. Note that even at such an early stage in a race as 30 feet (9.14 m) along the track, times can vary by nearly 0.2 sec between two athletes of comparable standard over 100 m.

Table I. Time measurements for the different athletes tested

	Time, sec	
	men	women
Rear block	0.123–0.182	0.132–0.225
Front block	0.327–0.370	0.339–0.430
Leaving blocks to 20 ft	0.907–1.000	1.041–1.058
20–30 ft	0.409–0.455	0.451–0.463
Gun to 30 ft, total	1.729–1.900	1.934–2.049

Impulse Measurements

It was hypothesised that the effectiveness of the start would be related to the average or mean acceleration on the blocks and, in particular, to the mean horizontal acceleration within the limits of reasonable balance in the first steps. In order to establish this parameter the impulse $\int Fdt$ was divided by the body mass, W, and the time, T, on the blocks, i.e. mean acceleration = $\int Fdt/WT$. Thus varying body masses and times on the blocks were accounted for. The relationship of the mean horizontal ac-

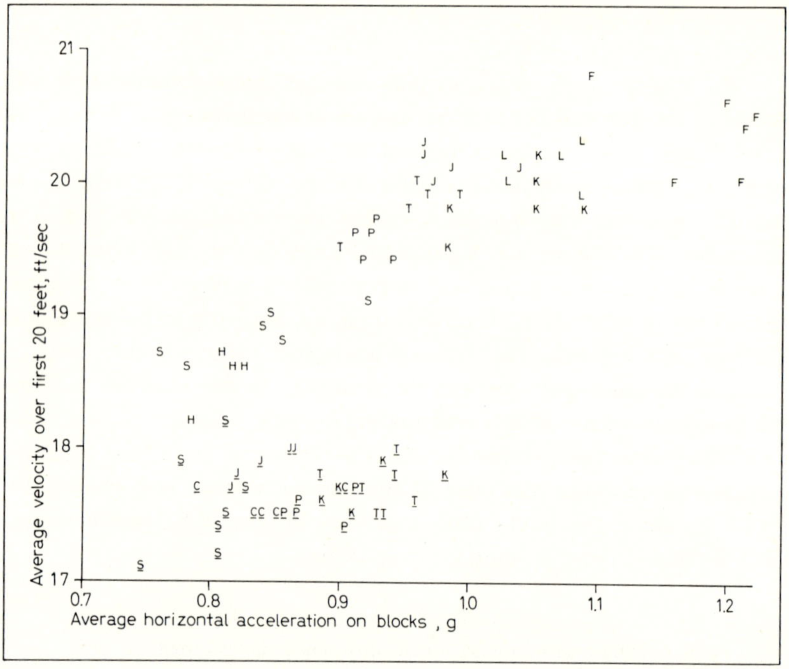

Fig. 4. Scatter diagram of athletes' mean velocity during first 20 feet of run against mean acceleration while on the blocks.

celeration and the mean velocity over 20 ft is shown in figure 4. There appears to be different tendencies in the relationship for men and for women, indicating that the women are not able to reduce substantially their times to 20 ft with increased acceleration on the blocks. (Plots of women's performances are shown underlined.) This may be due to anatomical differences, or it may be that women subjects in this study were nearer to peak efficiency than the men since it can be assumed that the relationship between acceleration on the blocks and this velocity will not continue linearly indefinitely, but will eventually flatten out. Thus, provided the athlete does not trip or interfere too much with the subsequent running, it would seem that on the whole as much horizontal impulse as possible should be striven for during the start. It was generally found that when athletes complained of 'stumbling' out of their blocks, they had the best starts as measured by the mean horizontal acceleration and mean velocity over 20 ft.

Acknowledgements

The authors wish to thank Mr. N. Mooney, Mr. S. K. Joshi, and Miss B. N. Knapp for their help with this project. The work was made possible by a Science Research Council grant.

References

Blader, F. B. and Payne, A. H.: Instrumented starting blocks. Athlet. Coach *2:* 3–4 (1967).

Blader, F. B.: The analysis of movements and forces in the sprint start. Biomechanics. I. 1st int. Sem., Zurich 1967, pp. 278–281 (Karger, Basel 1968).

Henry, F. M.: Force-time characteristics of the sprint start. Res. Quart. amer. Ass. Hlth phys. Educ. *23:* 301 (1952).

Authors' address: A. H. Payne and F. B. Blader, Physical Education Department, University of Birmingham, Edgbaston, *Birmingham 15* (England)

Medicine and Sport, vol. 6: Biomechanics II, pp. 232–235
(Karger, Basel 1971)

The Mechanics of Take-off in High Jump

R. Gombac

Jugoslav Institute for Physical Culture, Belgrade

The height of the take-off depends to a great extent on the physical and technical abilities of the jumper. In the relationship of these two factors, the level of physical abilities determines the jumper's possibilities and the level of his technique determines the degree to which his possibilities will be realized.

The aim of the analysis of the technique of take-off is to establish the economy of movements in relation to the level of physical possibilities, and to determine to what degree a rational technique can be acquired to cope with insufficiently developed physical abilities.

In order to obtain the necessary elements for analysis, the mechanical characteristics of the take-off should be determined first, as the physical and technical abilities of the jumper can be appraised e.g. on the basis of these characteristics.

To obtain an idea of the mechanical characteristics of take-off, we have registered over 60 jumps of 3 Jugoslav high jumpers, the best jumpers at the moment. They all used the Straddle style.

Each jump was recorded both by filming and by a dynamograph (the jumpers were taking off from a special tensiometric platform which allowed all changes in pressures to be graphically registered).

All dynamograms clearly showed the same characteristics. We shall discuss them in our report so far as the differences noted in the dynamograms show differences in their technical and physical possibilities. On the basis of films, kinograms were made, the centre of gravity was determined and the velocity of its movement established.

By analyzing the material under observation, we concluded that the take-off itself (run-up and bar clearance will not be considered here) may be divided into three phases as follows: placing the leg for take-off, amortization, active take-off.

1. Placing the Take-off Leg

This is the shortest phase, lasting about 0.020 to 0.024 sec. However, it is in this phase that the jumper exerts the greatest pressure on the ground (over 350 kg). The horizontal component of the pressure is considerably smaller and does not reach its highest value during this phase. Good jumpers always try to place their sole entirely on the ground and try not to drive their heels hard into the ground. It is hard to say whether this is at all possible. With our jumpers it is obviously not the case (see contour No. 2 in fig. 1 and phase I in fig. 2). They hit the ground hard with their heels, exert great pressure, and in the course of lowering their foot they let pressure fall to approximately 50 kg below the value of their bodyweight. The improvement of the technique in this phase should be directed towards shortening the time of putting the whole foot down, and maintaining a better contact with the ground, which will prevent such a great drop in pressure.

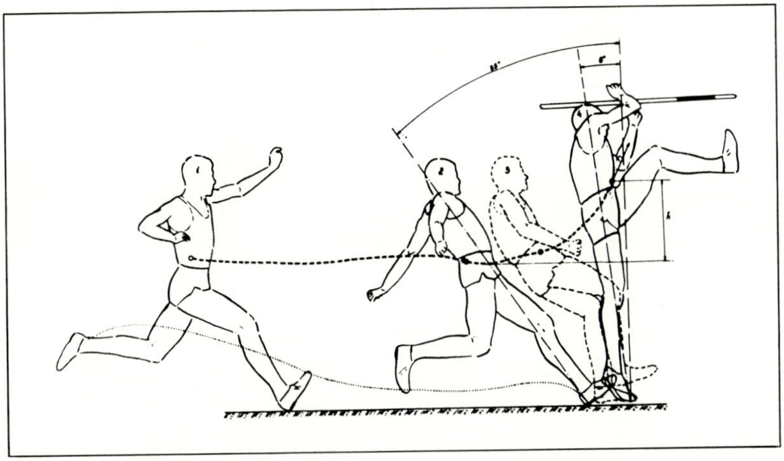

Fig. 1. Kinogram of the jumper. Contour No. 1 is the jumper in his last step but one; No. 2 shows the moment of placing his leg in take-off position; No. 3 is the end of the amortization phase, and No. 4 leaving the ground. The movement of the centre of gravity is marked by a dotted line. The difference in height of the centre of gravity counted from the moment of placing of the leg to its leaving the ground is indicated by 'h'.

2. *Amortization*

The amortization phase begins with the establishment of a firm support of the whole foot of the take-off leg.

In the case of our jumpers this may last between 0.10 to 0.12 sec, which is considerably longer than with the world's best jumpers. This extended phase primarily shows insufficient strength of the take-off leg.

Figure 2 shows that at the beginning of this phase the pressure increases very rapidly, *reaching a value almost identical with the one at placing the leg*. Immediately thereafter, because of heavier bending of the knee, the pressure goes down again by 100–150 kg and only when the bending of the knee has reached its extreme point does the pressure begin to increase again. The horizontal component of the pressure reaches its highest value at the beginning of this phase, when the vertical component has also reached the highest value.

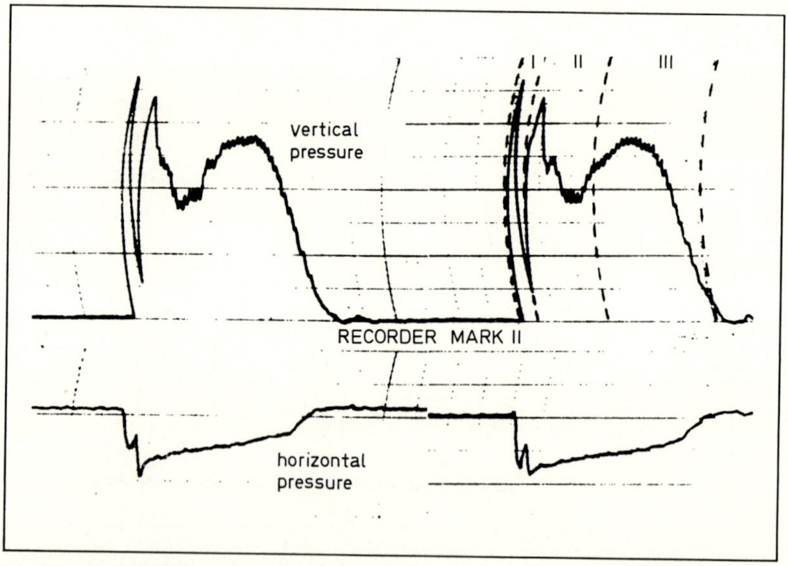

Fig. 2. The jumper's dynamogram. Both dynamograms were obtained from a recorder. The dotted lines in the second dynamogram show the difference between the various phases. Each horizontal line on the recorder's tape stands for a pressure of 10 kg. The upper part of the dynamogram represents the volume of vertical pressure, and the lower part the volume of the horizontal pressure.

The work on the improvement in this phase should be directed primarily to shortening of its duration which will also prevent excessive bending of the leg.

3. Active Take-off

This is the third and most important phase of the take-off. All earlier movements were made to create the most favourable conditions for this part of the work, the upward drive. The vertical component of the pressure rises again, but remains some 100 kg below what it was at the moment of placing the leg. The pressure produced by top jumpers in this phase reaches a value equal to the first phase or is just somewhat below it. The smaller force produced during this phase is a result of slow extension – in our jumpers this phase lasts about 0.125 to 0.150 sec – which is mostly due to insufficient force of the take-off leg and an unfavourable position from which the active take-off begins.

Author's address: Dr. RUDI GOMBAC, Jugoslav Institute for Physical Culture, Kneza Viseslava 72, *Belgrade 8* (Jugoslavia)

Medicine and Sport, vol. 6: Biomechanics II, pp. 236–239
(Karger, Basel 1971)

Kinetics of Pull-ups
Overgrasp vs Undergrasp

B. Ricci

Laboratory of Applied Physiology, Boyden Gymnasium, University of Massachusetts, Amherst, Mass.

I. Introduction

Pull-ups, variously called chinning, chin-ups, or chinning-the-bar, often comprise an important part of the evaluative portion of fitness tests for men. Existing pull-ups norms merely reflect numbers of completions relative to chronological age. The wisdom of this practice is seriously questioned. It seems almost unnecessary to point out that norms of this type are devoid of significance and utility.

The following example (table I) serves to clarify this point. Males A and B of similar age and height perform pull-ups. Vertical body displacement (measured at the suprasternal notch) amounts to 50 cm/pull-up.

Table I.

Subject	Weight kg	Pull-ups pu	pu/kg	Displacement m	Pos. work kgm
A	50	3	0.06	1.5	(50 × 1.5) = 75
B	100	6	0.06	3.0	(100 × 3.0) = 300
B	100	1.5	0.015	0.75	(100 × 0.75) = 75
B	100	3	0.03	1.5	(100 × 1.5) = 150

Note that subject B must perform *double* the number of pull-ups to achieve the *same* pu/kg body wt. index. Yet, when expressed as kgm work, B's accomplishment is *four times greater* than that of A. To attain the *same* work output, it is necessary for B (third entry) to perform *one-half* the number of pull-ups as A, which amounts to *one-fourth* the pu/kg index.

As portrayed by the last entry, the accomplishment of equal numbers of pull-ups (3) by persons of diametric mass (50 kg vs 100 kg) may provide justification for three varying statements: (1) that the individuals are *equal* in terms of number of pull-ups or (2) that *A* is to be ranked *higher* than B because of achieving a greater pu/kg index, or (3) that *B* is to be ranked *above* A because he performed twice as much work. This point need not be belabored. Suffice it to say that sweeping reforms are long overdue, in fact urgently required, to translate test items of so-called tests of 'physical fitness' into expressions which consider—*not totally disregard*—biomechanical factors [6].

II. Arthrological Aspects

Irrespective of the hand grasp utilized, elevating the chin to the level of the horizontal bar involves ginglymus action at the elbow joint, rotation at the proximal and distal articulations of the radio-ulnar joint as well as spheroidal action at the shoulder joint. Action at the wrist joint is dependent upon the grasp used. With the regular or overgrasp, pronounced extension at the wrist is observed when the chin is raised to bar level while with the reverse or undergrasp, pronounced flexion at the wrist is observed especially as the chin nears the level of the bar. The range of movement at the elbow joint is maximal; however, the shoulder joint range of movement is dependent upon the plane in which the elbows are moved. This movement, in turn, is dependent upon the amount of spacing between the hands on the bar as well as the displacement of the trunk and lower limbs.

A limited amount of motion may be detected within the sternoclavicular articulation. Displacement of the scapula is characterized by elevation: depression, upward-downward rotation, and abduction-adduction.

The aforementioned displacements, movements, and involvements were established by electrogoniometric and cinematographic monitoring.

Definition of Terms

Pull-ups describes the vertical displacement of the initially extended body until some reference point, notably the chin, is brought to the level of the wrists. A horizontal bar constitutes the usual, essential apparatus.

Overgrasp (or regular grasp) is descriptive of the hand grasp on the horizontal bar in which the performer's palms are directed *away* from his face.

Undergrasp (or reverse grasp) is characterized by palms being directed *toward* the performer's face.

Moments (MO) are descriptive of weight and inertial forces of each body segment.

III. Method

One 45-year-old male served as the subject for this study (table II).

Table II. Characteristics of subject

Subject	Sex	Age	Height cm	Weight kg	Body fat [1] %	Spec. gravity [2] kg/l
B.R.	♂	45	176	81.49	21	1.082

1 Technique of BEHNKE [1]
2 Archimedian principle, technique of BEHNKE *et al.* [2].

Photographs of the action were recorded at 32 frames/sec on 16-mm film with camera shutter one-fourth open. Both front and side views of the pull-ups were filmed. The action was projected frame by frame on a screen located at a distance from the projector which resulted in a scale of 4.5, i.e. 4.5 × tracing = life size. Joint centers were plotted from every third frame. Composite tracings of the total motion were then made and from these, segment displacement and length were calculated. Segment weight [1] and center of gravity [2] were also determined. These data were computerized according to the technique of PLAGENHOEF [4, 5].

IV. Results and Discussion

Results of pull-up performance were expressed in gram centimeters. Irrespective of grasp, the predominant forces in the execution of pull-ups are provided by the extensors of the humeri.

The contribution of the forearm flexors, by comparison with the arm extensors is secondary. This should not elicit surprise as muscle force is related to muscle cross-sectional area and moments at the elbow relate to moments at the shoulder. Cross-sectional areas of the arm extensors provided by the broad latissimus dorsi and posterior deltoid far exceed the cross-sectional area of the forearm flexors, notably the biceps brachii, brachialis anterior, and brachioradialis muscles. Scapulae stabilization by the trapezius and rhomboids is recognized.

1 Water displacement technique.
2 From DEMPSTER [3].

Comparison studies of the effectiveness of overgrasp vs undergrasp on pull-up performance are noticeably absent from the literature.

Substantiation for the greater accomplishment with undergrasp was rendered understandable by focusing attention on the moments at shoulder, elbow, and wrist joints. The minus signs (in overgrasp) in each of the positions except starting and end position reflects dominant extensor action of the shoulder joints. With undergrasp, a change in force magnitude and direction was observable at the aforementioned joints, and was indicated by the change in sign at varying positions of the wrist and elbow joints. Thus, muscle force application while utilizing the undergrasp is not of a sustained nature. Consequently, overlying forces on blood vessels (vis a latere) would produce less disruptive effects relative to blood delivery to active muscle tissue including the removal of end-products of muscle metabolism. Greater and sustained involvement of the forearm flexors with overgrasp was indicated by larger minus values and larger moments in comparison with undergrasp.

The subject for this study was unusual because of complete lumbar fusion. This may explain, in part, the reason for the pronounced forward body movement as the pull-up neared completion.

References

1 Behnke, A.R.: Quantitative assessment of body build. J. appl. Physiol. *16:* 960–968 (1961).

2 Behnke, A.R.; Feen, B.G., and Welham, W.C.: The specific gravity of healthy men. Body build, volume as an index of obesity. J. amer. med. Ass. 498 (1942).

3 Dempster, W.T.: Space requirements of the seated operators; geometrical, kinematic, and mechanical aspects of the body with special reference to the limbs. WADC Technical Report No. 55–159 (Wright Air Development Center, ARDC, USAF, 1955).

4 Plagenhoef, S.C.: Methods for obtaining kinetic data to analyze human motions. Res. Quart. amer. Ass. Hlth phys. Educ. *37:* 103–112 (1966).

5 Plagenhoef, S.C.: Computer programs for obtaining kinetic data on human movement. J. Biomech. *50:* 221–234 (1968).

6 Ricci, B.: Experiments in the physiology of human performance (Lea & Febiger, Philadelphia 1970).

Author's address: Prof. Benjamin Ricci, Laboratory of Applied Physiology, Boyden Gymnasium, University of Massachusetts, *Amherst, MA 01002* (USA)

Medicine and Sport, vol. 6: Biomechanics II, pp. 240–245
(Karger, Basel 1971)

The Phenomenon of the Freedom of Lateral Deviation in Throwing (Wurfseitenfreiheit)

E. LINDNER

Institut für Leibesübungen der Universität Marburg, Marburg a. d. Lahn

This report presents an analysis of the release phase of the following subjects: German 'Schlagball', baseball, javelin, German handball, and shot put. Such an analysis constitutes a central problem of modern throwing techniques.

Owing to limited space we are unable to present the entire development of our experiments here. It is possible only to outline the results briefly.

It must be premised that until now we have observed movement processes from a certain position, primarily from the side. Moreover, literature has helped to determine certain conceptions of movement. Therefore, it is particularly interesting and worthwhile to observe movements from above, from behind, and from the front of a subject, because from such positions the final phase of the throw can be observed most conveniently. Also, a 'many-sided' analysis supports the conclusions.

Our aim is to reduce prejudices which have led to false doctrines based upon misleading theoretical directions and implanted conceptions of movement. From literature as well as from discussions with coaches and athletes the following conceptions of movement have emerged.

Towards the final throwing phase the arm moves to a position close to the head and, if possible, above the shoulder. Only then is it possible to develop a push and reduce the danger of injury to a minimum.

The direction of the movement as made from the start is determined by the trace of a run or slide and is followed through the final phase by the whole body.

These conceptions of movement are incorrect and yet so widely accepted that even athletes who are regarded as specialists in throwing seriously, believe that the arm performs an actual and releasing jerk above the shoulder and near the head. Medical literature gives instructions calling for a release movement near the head by explaining that the danger of injury increases with the distance of the action from the head and the shoulder. We do not condemn this opinion, but we compare it with modern throwing technique observed in untrained young people as well as in top athletes. The arm movement is made at the almost greatest distance from the head. Our statement is particularly relevant in so far as we have found this peculiar phenomenon in several athletic throwing sports. Therefore, we believe it to be a dominant factor of the normal shape of movement of these disciplines.

First of all, let us consider the movement of the throwing arm. It can be characterized as follows. Supported by a tense bending and turning of the trunk, the arm moves forward and upward from a low, backward starting position. The path of the movement is not plane but spiral. An examination of this basic movement shows these results.

1. German 'Schlagball'

The pitch in 'Schlagball' is typically a flinging movement allowing ample freedom of movement for the flinging arm, i.e. a lateral movement not limited by the object to be thrown. The movement as seen from above shows [LINDNER, 1969] that the flinging movement of the arm takes place on the side of the body (fig. 1). A lateral deviation of the upper part of the body with respect to the direction of running is typical of this movement. We suggest to call this phenomenon *'Freedom of lateral deviation in throwing'* («Wurfseitenfreiheit»). In Germany, this has become a standard term. It seems to be a suitable term for athletes and coaches seeking a correct description of movement.

2. Baseball

The same phenomenon can be noticed in the baseball pitch. Arm and shoulder reach almost a straight line in the final phase. Front-view and back-view observations illustrate clearly the typical position of arm and shoulder in the release phase (fig. 2 and 3). The arm is straight and in line with the shoulder's axis as the object is jerked away.

Fig. 1. 'Schlagball'

Fig. 2. Baseball

Fig. 3. Baseball

3. Javelin

In the final phase many athletes step sidewards in the opposite direction of their throwing arm. The wide track of running and the sideward movement of the upper part of the body allow free development of the throwing phase as well as a good transfer of energy of the whole body to the flinging arm ('Armschleuder'). Thus, freedom of lateral deviation of the throwing arm is created. From an intense backward bent position of the body the arm and the shoulder are jerked forward, a movement in which the throwing arm performs a flinging motion *beside the body and not above the shoulder*. With respect to the part of the shoulder near the joint with the throwing arm, the javelin is jerked away from a lateral arm position (fig. 4, 5, 6). The release position 'above the shoulder' or even 'above the head', as stated in the literature, cannot be found in any top athlete [LINDNER, 1969].

A body moving at a velocity 'v', suddenly meeting with an eccentric resistance, will turn round the point where the resistance takes hold of it. The points farthest from the rotation centre will, compared with 'v', move at a higher speed. We mention this phenomenon since it supports our theory of the lateral positioning of the left foot with its stopping function and of the freedom of lateral deviation of the right arm with its flinging function at the right side of the body (fig. 5, 6).

The bending of the upper part of the body to the left is also advantageous in order to control and balance one's weight, since owing to the

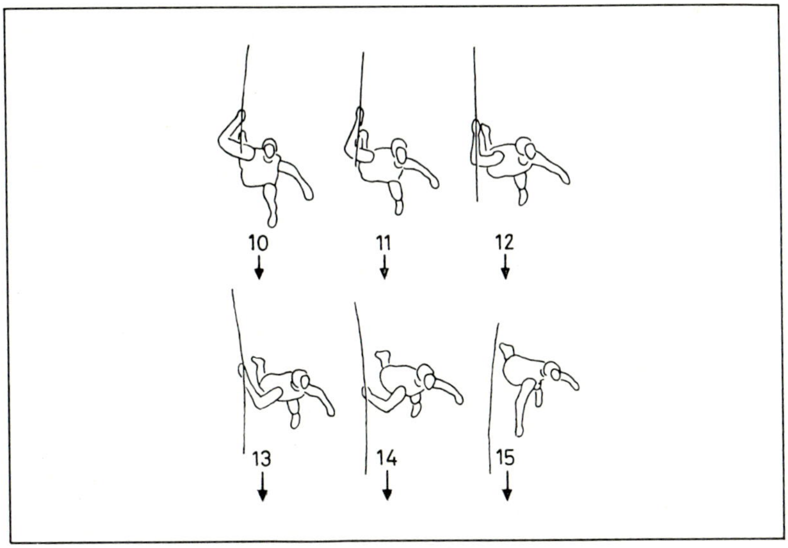

Fig. 4. Javelin from above.

Fig. 5. Javelin (Lusis).

Fig. 6. Javelin (Kulstar).

Fig. 7. Handball

Fig. 8. Shot put

inertia of the body, the force acts in the direction of the stemming left leg and is thus balanced. The head and the upper part of the body bend towards the left. Through this bending of the body, the body tension affects the throwing movement of the arm favourably.

4. German Handball

The release phase of throwing a handball resembles closely that of the javelin throw as can be observed from the final step sideward and the jerking phase. As the upper part of the body bends in lateral deviation to the left, the arm on the right side of the body flings forwards and straightens (fig. 7).

5. Shot Put

One would think that owing to the heavy weight of the shot, the weight of the body is directly behind the object to be thrown. It is, therefore, all the more surprising that in the final phase the same phenomena occur as have already been mentioned. As figure 8 shows, the shot is swung to the side of the putting direction.

6. Conclusion

It is evident that lateral deviation with a powerful trunk and shoulder action supports the power output of the moving arm. Moreover, since this phenomenon has been observed not only in trained experts but also in children, it must be considered a natural movement. It is possible that our findings will be of value for future basic and advanced training methods.

Reference

LINDNER, E.: Sprung und Wurf. Beiträge zur Lehre und Forschung der Leibeserziehung, vol. 29 (Hofmann, Schondorf b. Stuttgart 1969).

Author's address: Dr. ERICH LINDNER, Institut für Leibesübungen der Universität Marburg, Barfüsserstrasse 1, *D-355 Marburg a. d. Lahn* (Germany)

Medicine and Sport, vol. 6: Biomechanics II, pp. 246–248
(Karger, Basel 1971)

Changes in the Knee Angle and in the Load of the Ski during Swing Motions in Alpine Skiing

T. Fukuoka

Institut für Leibesübungen der Johann-Wolfgang-Goethe-Universität, Frankfurt a. M.

Changes in pressure between a ski and the snow surface and in the knee angle are two of the most essential conditions for effectuating a swing in alpine skiing (stem, parallel, and short swing).

The aim of our experiment is to observe these phenomena by an electronic measuring method which does not hinder the subject.

Method

It was necessary to develop a special dynamometer and goniometer for our experiment. For measuring we used a telemetric system, because its small dimensions are an advantage in data processing.

Dynamometer and Goniometer

Wire strain gauges in a bridge circuit were applied as a force transducer or dynamometer. The goniometer measures the angle between upper and lower leg and consists of a potentiometer bridge circuit. For attaching the goniometer to both sides of the subject's leg we used electrode bands as are normally used in EMG measuring.

Telemetric System

The Tetronik Co. Ltd of Wiesbaden made the telemetric system in accordance with our specifications. The telemeter system has 4 channels (2 for the dynamometer and 2 for the goniometer).

The small transmitter weighed about 1 kg and was attached to the subject's back by means of a special belt. A receiver converted the four output signals into a signal appropriate to an ultraviolet recorder with small galvanometers.

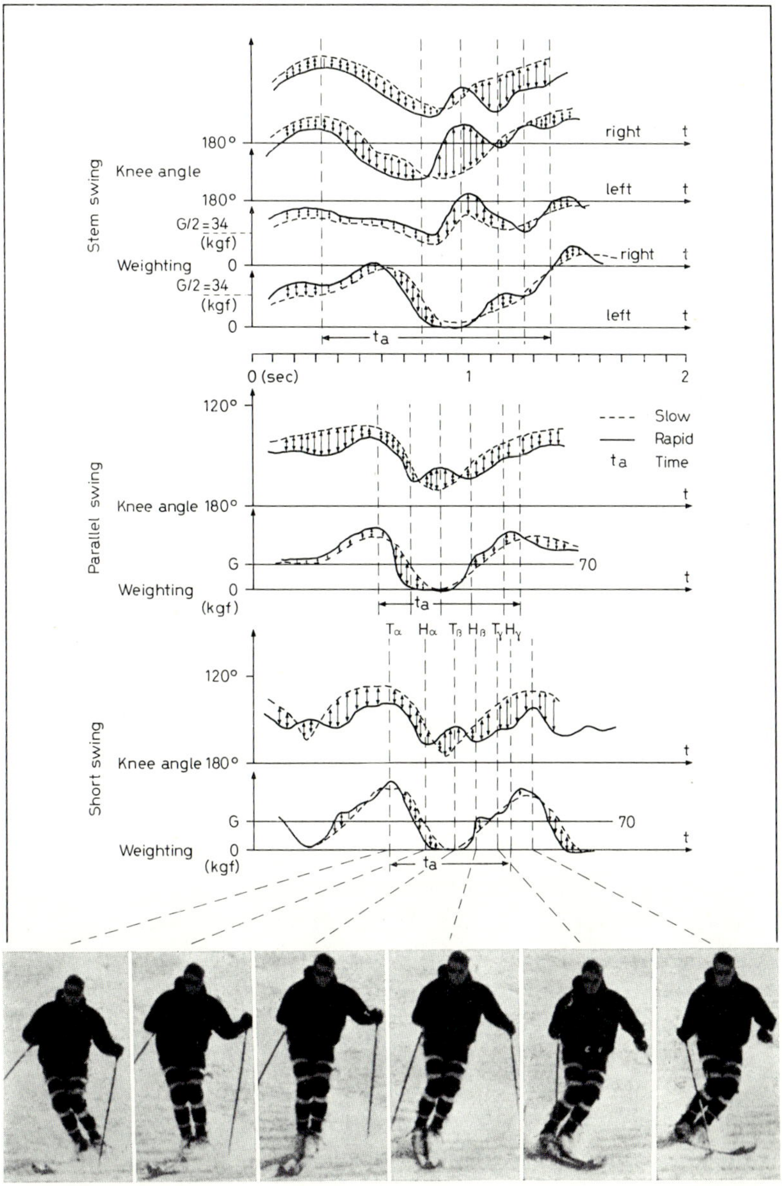

Fig. 1. Standard curves.

High-Speed Camera

For synchronizing the functional morphological motions of the subject and the measured signals we used a high-speed camera provided with a speed controller. The speed during the experiments was 80–150 frames per second.

The Experiment

Place of experiment: St. Christoph, Arlberg; Riezlern, Kl. Walsertal
Test persons: Skiers with different levels of training.
Number of tests: 50 test series of 5 trials each

Main Result and Discussion

Regarding the alterations in the pressure and knee angle the standard curve of each swing (stem, parallel, and short swings), which was ascertained through superposition, shows a characteristic pattern (especially more or less pressure phase, maximum pressure, angle velocity of bend and stretch motion in the knee joints and amplitude in the knee angle).

The characteristic points of changing in the knee angle pattern are:
Tα: first bend in knee joint (last phase of preceding swing with edging)
Hα: first stretch (push off from the snow surface)
Tβ: second bend (pulling up of the legs)
Hβ: second stretch (for turning the skis)
Tγ: third bend (landing on the snow surface)
Hγ: third stretch (compensation motion of the legs)

While swinging at a slow speed the standard curve of the knee angle shows only one point of the bend and the stretch. With increasing speed the bend and stretch points (Tα, Tβ, Tγ, Hα, Hβ, Hγ) increase.

Depending on the motion, there exist for each short swing two standard curves: (1) standard curve with weight-shifting, (2) standard curve without weight-shifting.

Both motions show one complete unloaded phase.

According to our experimental observations the consideration of knee angle and pressure patterns – as a consequence of reactional necessity for a swing in alpine skiing – is one of the most essential factors in classifying a training level of each swing method.

Author's address: TAKAZUMI FUKUOKA, Dipl.-Phys., Institut für Leibesübungen der Johann-Wolfgang-Goethe-Universität, Frauenlobstrasse 5, *Frankfurt a.M.* (Germany)

Medicine and Sport, vol. 6: Biomechanics II, pp. 249–252
(Karger, Basel 1971)

Biomechanics of Rowing

T. ISHIKO

School of Health Sciences, University of Tokyo, Tokyo

Introduction

Because rowing is performed on water away from the shore, analysis of the rowing movement is difficult. Movie technique has been frequently utilized, but this technique does not give us sufficient information, because the movement of the oars in the water and of the legs of the oarsmen in the boat are invisible from the shore. The author, therefore, applied telemetry to the study of rowing in order to obtain more knowledge [ISHIKO, 1968].

Method

Measurements were performed on the strain of the oar and on the acceleration of the boat. The strain of the oar was recorded by fitting 4 strain gauges on the on-board part of the experimental oar and by connecting them as a Wheatstone bridge. The electric current due to strain was amplified with the aid of a DC amplifier and transmitted by an FM transmitter to the recorded set-up on the shore. The strain was calibrated by hanging various weights at the grip perpendicularly to the oar. This calibration procedure enabled us to determine muscular force exerted perpendicularly on the blade at the grip of the oar. Acceleration of the boat was measured by using an accelerometer (strain-gauge device) placed on the bottom of the boat and recorded with the aid of a telemeter system.

Results

1. Forces at the Oar

The curves in figure 1 show the force perpendicular to the oar. From A to B the curve shows a slight increase. Comparing the curve with the acceleration curve of figure 3, part AB could not be produced by rowing

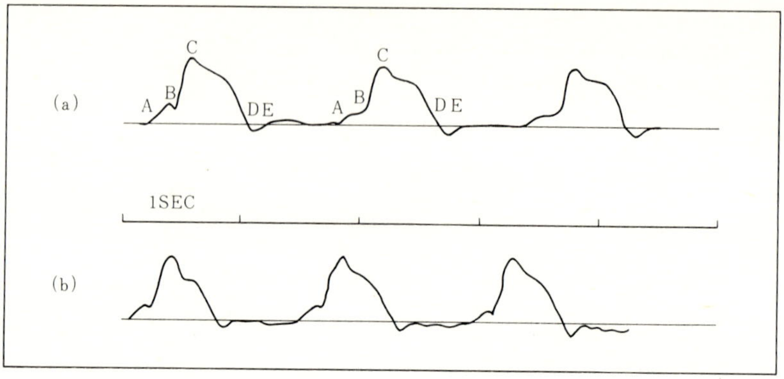

Fig. 1. Forces measured at the oar at a rowing frequency of 38 strokes (a) and 42 strokes (b) per minute.

in the water, because during the time interval AB acceleration was decreasing. Therefore, it may be concluded that it is due to the inertia of the oar and was generated by changing its movement from bowward to sternward direction. The duration of interval AB was about 0.2 sec, which was used to obtain a speed for the blade of the oar to equal that of the water flowing sternwards. Part BCD of the force-time curve shows the muscular force exerted on the oar. Counted from B, the curve increases rapidly to its maximum C, which corresponds to 80–100 kg. After this the curve decreases slowly to D, where no force is measured because the blade of the oar leaves the water. Curve DE shows a negative force due to the bending of the blade in the opposite direction caused by a sudden drop in the water resistance at the end of the rowing cycle in the water. There is hardly any force present during the interval E to A, where the blade is moved bowwards above the water to the position for the next stroke.

Figure 1a and figure 1b show force-time curves at the oar during a rowing frequency of 38 and 42 strokes per minute by the same subject. It is noticeable that the pause EA becomes only somewhat shorter as well as the duration of the energetic part of the rowing cycle when increasing the pace of rowing. However, the curve shows the same pattern. This pattern should be firmly established owing to the long period of experience in rowing and will not change even when the pace of rowing is altered.

Figure 2 shows a number of force-time patterns obtained by telemetry. The force-time curves vary according to the subject: each oarsman rows in his own way. Some oarsmen row in such a way that the force exerted

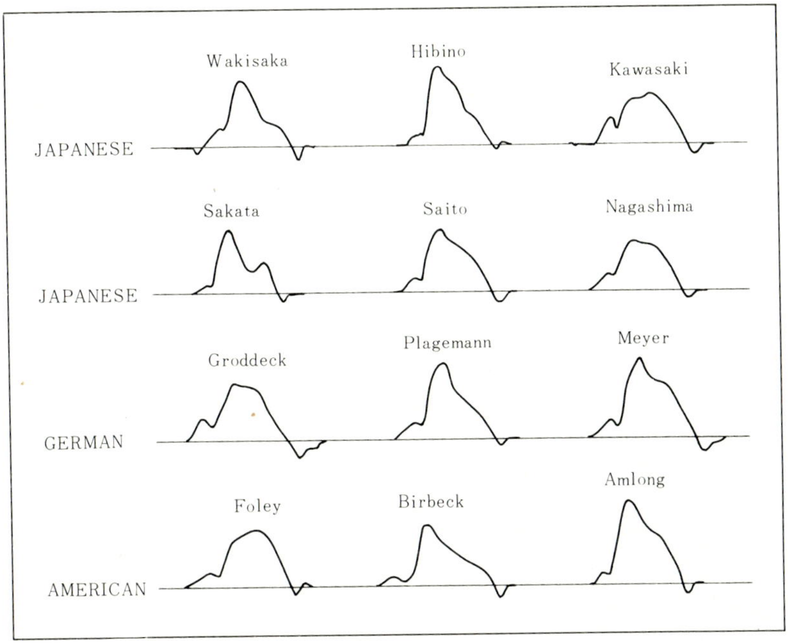

Fig. 2. Examples of force-time curves measured at the oar, by Japanese, German and American oarsmen.

reaches its maximum value soon after the blade has touched the water, while others reach their maximum force rather later. Some curves show a steep peak, others have more or less a plateau. No particular national or international difference in the characteristics of the force-time curves can be observed.

2. Acceleration of the Boat

Figure 3 shows simultaneous recordings of force at the oar and acceleration of the boat. The acceleration curve increases rapidly as soon as the oar touches the water. Maximum acceleration was obtained when the force reached its maximum. After that, the acceleration curve decreased with some undulations to zero. At this moment the oar left the water. Then acceleration increased again, although the oar was already above the water. This part of the acceleration is not due to the rowing force exerted but to the counter-movement against the body moving forward/sternward

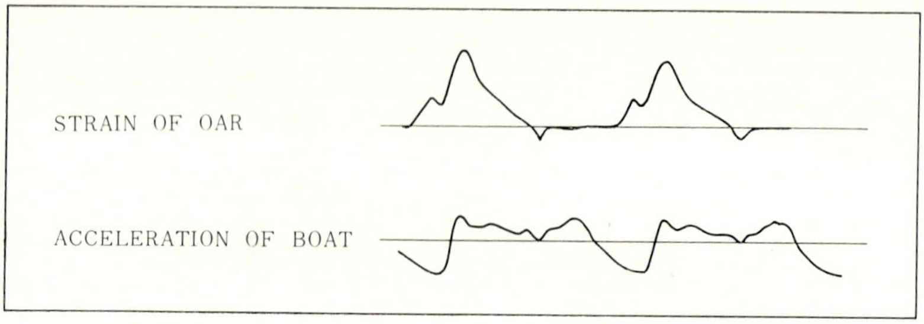

Fig. 3. Curves showing force exerted on the oar and acceleration of the boat.

on a sliding seat in order to be prepared for the next stroke. This increase is followed by a rapid decrease which is due to the water resistance. The amplitude of the acceleration obtained by shell boats is about 1.8 g.

Discussion

It was astonishing that the force-time pattern during rowing was different from one subject to another. This is true even among world-famous crews. The movement of oars of a crew above the water is nearly uniform, but this does not ensure uniformity of the movement of the oars in the water. Oarsmen of a crew try to row in the same manner and they believe they are doing so. But from the data it may be concluded that this actually is not true. Rowing looks very simple, but it is one of the most difficult sports, since oarsmen cannot know how they are really rowing. On account of the facts mentioned above we may suppose that 'feed back' of the movement involved may be a very important factor in motor learning in rowing.

It may be concluded that the recorded forces acting on the oar and acceleration of the boat give good information of the way of rowing and seem to give useful criteria for evaluating the rowing movement of an oarsman and a crew.

Reference

ISHIKO, T.: Application of telemetry to sports activities. Biomechanics. Proc. 1st int. Seminar, Zürich 1967, pp. 138–146 (Karger, Basel 1968).

Author's address: Dr. T. ISHIKO, 7-20-1 Nishi-Gotanda, Shinagawa-Ku, *Tokyo* (Japan)

Medicine and Sport, vol. 6: Biomechanics II, pp. 253–257
(Karger, Basel 1971)

Graphic and Electromyographic Analysis of Various Styles of Swimming

L. LEWILLIE

Free University of Brussels, Stress Laboratory, Brussels

The electrical activity of various superficial muscles was studied by means of surface electrodes according to the telemetric method presented at the 1st Seminar of Biomechanics [LEWILLIE, 1968]. Various technical improvements have been introduced to make the equipment more compact. Standardization was obtained by the incorporation of a generator which supplied a signal of 500 μV. Recording took place on magnetic tape. An integrating technique used during transcription made a more precise analysis of the results possible. A technique of luminous traces of the movements enabled us to compare the various styles as well as the various adaptations revealed during one and the same movement.

The electromyograms show with sufficient accuracy the differences in activity of a specific muscle which depends on the actual style of swimming, and also the individual differences which appeared even among swimmers of comparable proficiency, as well as the characteristics of the various speeds of swimming of one and the same swimmer.

The action of the extensor muscles was found to be essential. The muscles most characteristic of swimming are the m. triceps of the arm and the m. quadriceps of the thigh. The activity of the m. brachialis triceps lasts short and is intense in breast strokes, but its period of activity is very long in crawl. In the latter case one can see the appearance of important variations of the style, depending on the degree of proficiency of the subject, thus proving that a style of swimming, generally considered as natural, requires long training to obtain a normal efficiency.

Comparative examination of the electromyogram of the muscles of the upper extremity, as well as of the luminous traces, show the important contribution of these muscles in the butterfly style for the specific sweeping along of the swimmer. In one and the same swimmer the regularity of the movement is such that certain authors have used it as a criterion of the

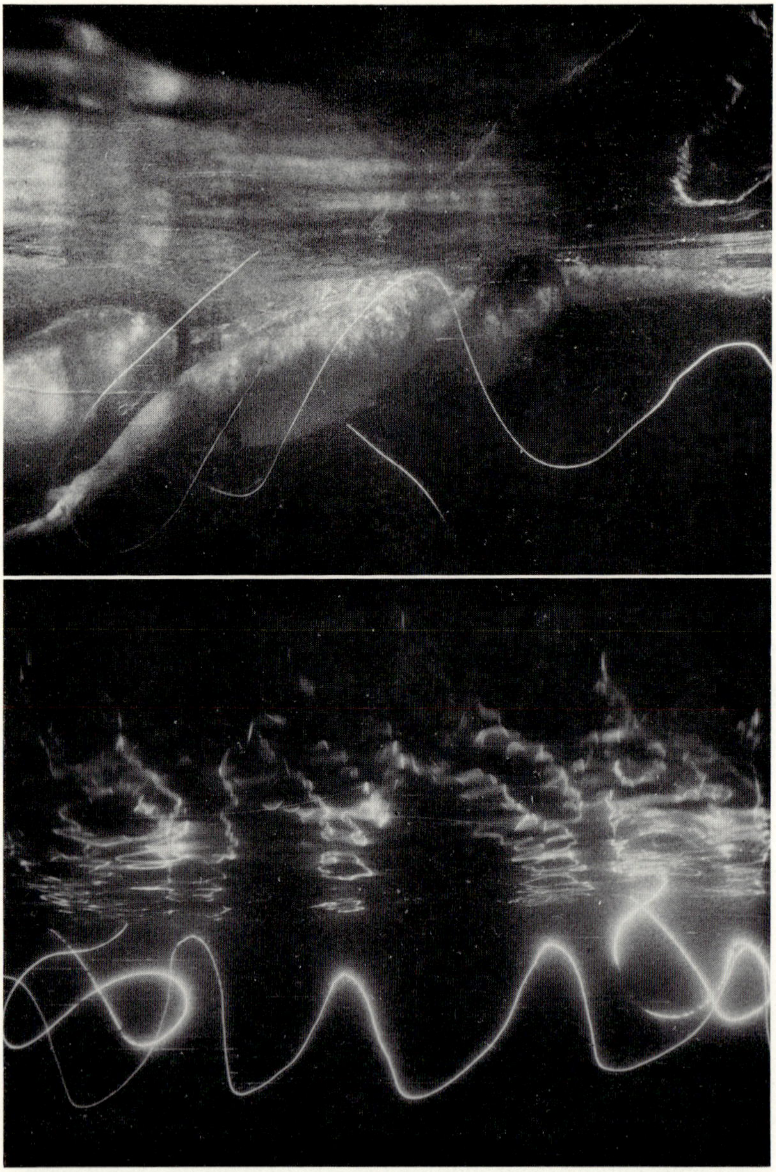

a

b

Fig. 1. Recording of the traces of the hand and the foot in four swimming styles. The subject is moving from left to right. *A* Crawl, *B* back stroke, *C* breast stroke, *D* butterfly.

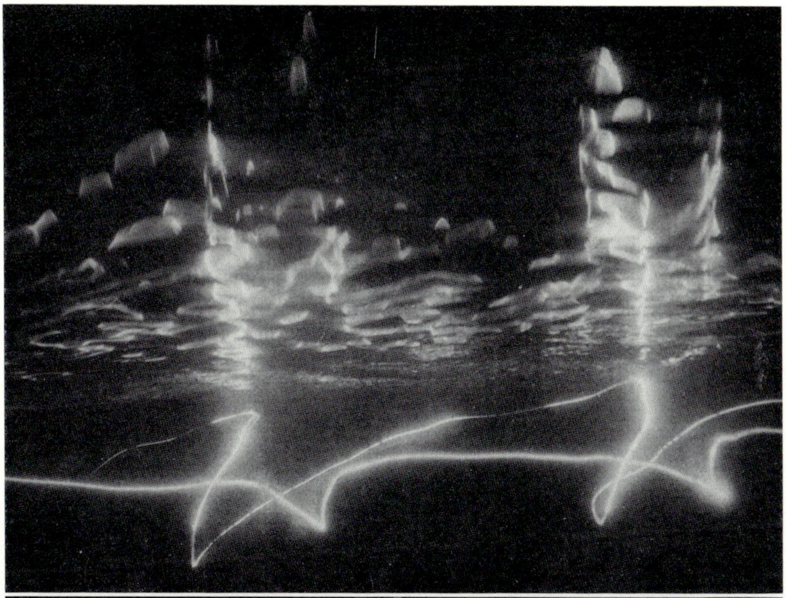

c

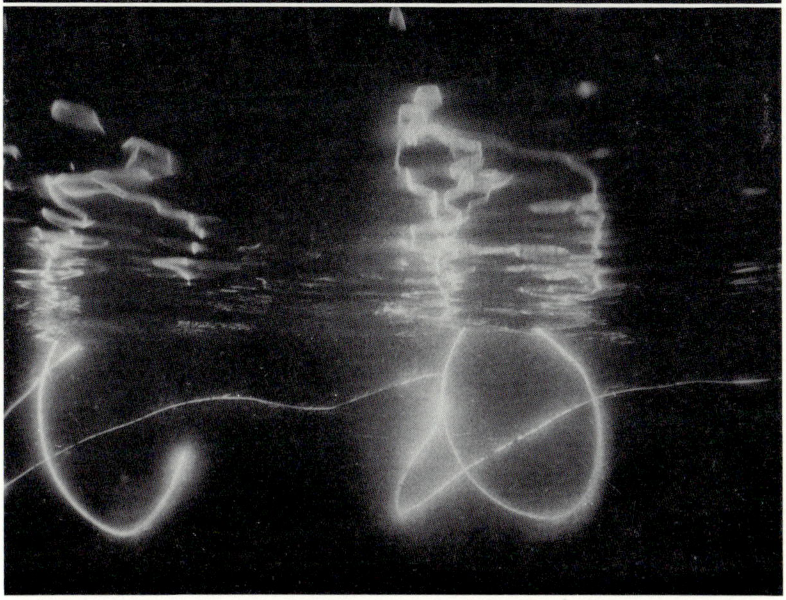

d

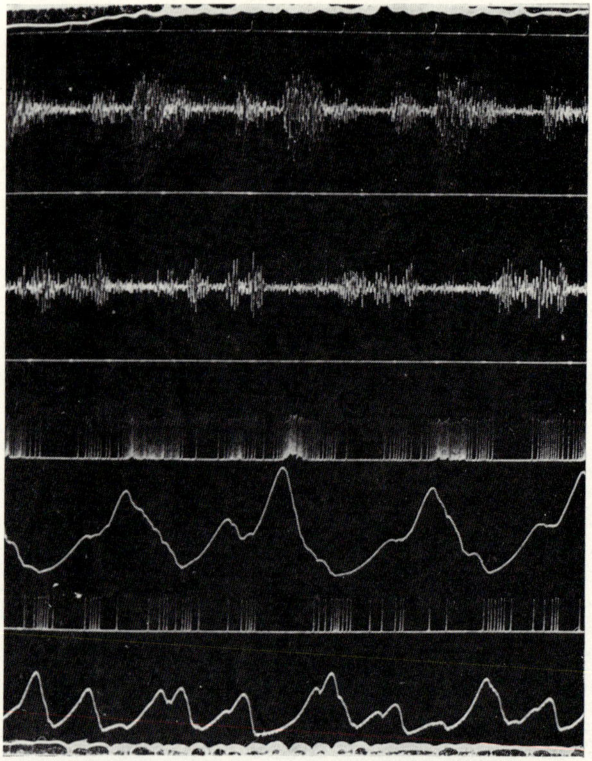

Fig. 2. Recording of the electromyogram of swimmer during crawling. *A* triceps of the arm, *B* and *C* integration, *D* quadriceps of the thigh, *E* and *F* integration.

speed. This is confirmed also by the duration and the amount of activity of the electrical phenomena of the muscle, but in the course of one cycle, which may cover several movements, it is, just as in the beat of the crawl, strongly influenced also by the respiratory movements.

The muscles not concerned directly with propulsion but influenced by a number of factors, such as attachment of the limbs, respiration, position of the head, etc., show a less pronounced regularity.

The difference in activity of the various heads of one and the same muscle show how carefully the non-specific development of a swimmer's muscles should be undertaken. A detailed analysis of the electromyograms recorded with the four styles of swimming is given.

The simultaneous analysis of muscles considered agonist and antagonist, permits the study of the effectiveness of the movements of various body segments.

Reference

LEWILLIE, L.: Telemetrical analysis of the electromyogram. Biomechanics. Medicine and Sport, vol. 2, pp. 147–149 (Karger, Basel 1968).

Author's address: Dr. LÉON LEWILLIE, Free University of Brussels, Stress Laboratory, *Brussels 5* (Belgium)

V. Research in Rehabilitation

Medicine and Sport, vol. 6: Biomechanics II, pp. 260–265
(Karger, Basel 1971)

Biomechanics of the Normal and the Pathological Gait

F. van Faassen, N. H. Molen, and W. Boon

Institute for Biomechanics and Rehabilitation, Free University, Amsterdam

Introduction

At the Institute for Biomechanics and Rehabilitation of the Free University in Amsterdam, research has been done regarding the patterns of movement displayed by physically normal subjects as well as by those who are in one way or another handicapped. In order to determine these patterns in both groups, a number of measuring methods were developed, while other methods already known were modified in such a way that they, too, could be used within this special research program.

When applied in the case of physically normal people, these methods enable us to determine objectively their moving patterns in measures and numbers; applied in the case of those who are handicapped, they give us an idea of how their movements differ from those of their bodily healthy counterparts. Through periodic measurement of the handicapped we can assess to what extent the therapy prescribed is successful.

The measuring takes place on a walkway or on a treadmill. The rubbery surface of the former is conductive, because it is prepared with carbon. It is approximately 25 m long. The length, of course, is necessary to eliminate from the recording both the start and the finish of the actual experiment; the remaining length is sufficient to get statistically valid results. The treadmill is 6 m long, 80 cm wide, and allows of being put in a sloping position varying from 0–9 °. Its speed, which can be controlled without interruptions, varies from 0–240 m/min. In what follows the methods now in use are briefly described.

I. Measurement of the Temporal Factors of the Stride

Walking consists of a cycle of movements repeating themselves continuously. When recording a subject's pattern of movements it is, therefore, desirable to consider this pattern as a function of the time. As is well known, the stride can be analyzed to various contributing temporal factors such as stance time, swinging time, double support time, and the step time of either leg.

The step-time signal is used as basic time for all other experiments directly related to the measuring of movements. In this signal, all individual factors are combined so as to form a timing diagram in such a way that they can easily be read separately. The recording of the walking pattern is by means of conductive soles, attached to the shoes.

Contact and interruption of contact with the carbon-prepared surface of the walkway and/or treadmill records the temporal factors of the stride on one channel of the recorder.

The factors constituting the temporal factors are in themselves already valuable sources of information with respect to the manner in which a person walks; therefore, they are not only used as basic time for other measuring experiments, but also studied for the sake of their own absolute value.

In order to obtain rapidly a convenient survey of the pattern of the temporal factors, all time factors to be distinguished are automatically typed in tabular form by means of a digital data device. Included in the tabulation is the average value of the quantities measured from an adjustable number of steps. For statistical analysis a punched tape is made.

II. Stick Diagrams

In order to be informed about geometric patterns of movement of the lower extremities, stick diagrams are made.

The subject is dressed in black leotards. The parts of the body to be investigated, that is the upper and lower legs, are each provided lengthwise with a white, light-reflecting tape. The knee and the nose of the shoe are marked with a cross. Thus dressed, the subject walks on the walkway and passes a technical camera, the shutter of which is open. In front of the lens of the camera there is a rotating disc with a slit, so that during every rotation the photographic plate in the camera is exposed for a very short time (0.001 sec). Since only the white tapes reflect, and the background is pitch dark, the position of the body parts under observation are photographically recorded every time the disc rotates, which is 24 times per second. As points of reference there are some white vertical lines hanging in front of the black background at a fixed distance from each other. For the sake of time synchronization, electric pulsations are produced simultaneously with the flashes of the area; they are recorded together with other quantities.

Recordings thus obtained provide us with information regarding the position of the extremities relative to the vertical lines as a function of time. From this position we learn the angle between upper leg and vertical line and that between upper leg and lower leg, as well as the various angular speeds. Finally, the stick diagrams furnish information regarding, among others, the forward speed, the maximum elevation of the heel, the length of the step and displacement of the centre of gravity.

III. Measurement of the Knee Flexion Angle during Walking

Since it is rather time-consuming to work out the stick diagrams and also since the recording is restricted to one or two steps, it is advisable to measure some quantities separately, and from a greater number of steps. Thus the knee flexion, for example, although clearly indicated in the stick diagram, is separately recorded by electric knee-flexion transducers.

They are attached to the knees in such a way that their rotation axes coincide with the axes of the knee joints. The deflections given by the transducers are calibrated by placing the knees in two different positions, for instance, at angles of 45 and 90 °. During walking the positions are measured by the device and from an arbitrary number of steps the knee flexion as a function of time is then determined.

IV. Measurement of the Step-length

The step-length, too, can be measured on the stick diagram. However, here too applies that the measurement covers one or two steps only. Since statistics require a greater number of steps, another method of measuring is used to determine averages and standard deviations. This also makes it possible to express quantitatively the geometric asymmetry of the step-length. The step-length is measured in combination with the measuring of the temporal factors. For this purpose, the subject walks over the prepared track on special soles. During each double support time the step-length is digitally recorded.

V. Measurement of the Momentary Velocity

In trying to come to valid conclusions regarding the walking pattern, a person's speed is an important factor. The walkway is, therefore, equipped with a measuring device. At any given moment the speed of the centre of gravity of the body in forward-backward movement can be measured in a simple but reliable way. In fact, while walking, the subject moves a magnetic tape attached to his body. The signal on the tape is detected by a magnetic pick-up. The signal frequency is proportional to the momentary speed of the subject. From this frequency-modulated signal and the momentary speed, the accelerations and decelerations are recorded by means of an analogue computer, while the average speed over a given distance is also directly readable. The acceleration curve thus recorded gives some information about the external forces that act on the centre of gravity of the body.

For more accurate information on this point the reaction forces of the feet have been measured as well.

VI. Measurement of the Reaction Forces of the Feet

This measurement, which can only be done on two steps, is three-dimensional. While walking on the walkway, the subject steps on two force-measuring platforms invisible to him. They are installed in such a way that the left and the right foot step on the left and the right platform respectively, with the result that the reaction force of either foot is measured separately in three mutually perpendicular directions. In this measuring experiment the weight of the body and the speed of walking are important factors, to be properly taken into account when the results are worked out.

There are a great number of variables in the walking pattern. Therefore, it may be necessary to fix some of the parameters when determining norms or testing results to these norms. For example the step frequency, i.e. the number of steps per minute, can be dictated with an acoustic signal of an electronic metronome. The danger is, however, that a possible asymmetry of the time in the walking pattern is influenced by the rhythm of the metronome. It is also possible to obtain the mean walking speed of a subject by means of a mechanical 'hare'.

The speed of the 'hare' can be chosen independently of the subject. The subject following the 'hare' has, of course, the same average speed. Step frequency and speed being fixed, the average step-length is fixed as well. However, it is also possible to dictate separately the step-length with the help of marked points on the surface of the walkway.

Since, as a result of the above methods the subject has to concentrate on the way he walks, this may influence his walking pattern.

For the kinematic-dynamic research, a 'treadmill' is available. In addition to the above measurements several others are performed. The temporal factors and the knee flexion, for example, are measured in the same manner as is done on the walkway. Like the walkway, the track of the treadmill is conductive. In order to make stick diagrams special arrangements must be made, because the subject does not actually move face to face with his surroundings, which would result in pictures covering each other and making observations impossible. In order to prevent this, a so-called cycle choice method is necessary. This method entails that instead of one stride, twice half a stride is photographed, and the chosen cycle is synchronized with the measuring of the temporal factors of the stride. The shutter of the camera is opened and closed as a function of the pattern of the temporal factors. On the treadmill only the average values of the speed and the step-length are determined. The speed of the treadmill is adjusted arbitrarily and independent of the subject's rate of walking while the average step-length is calculated from the average speed and time. Since the treadmill cannot be equipped with force measuring platforms, approximately the same results as in the case of platforms can be obtained by equipping the shoes with pressure transducers. It is difficult to reproduce this method because it only records local pressure between foot and shoe; it is, therefore, applied only as a comparative measurement.

VII. Measurement of Energy Expenditure

This measurement, too, is performed on the treadmill. While the subject walks on it in accordance with its speed, the oxygen consumption and the carbon dioxyde production are measured by means of the indirect open method with a Noyon's diaferometer. A two-point recorder gives the results worked out by a computer for statistical analysis. Next to this method there are other methods available in order to determine oxygen consumption. Other physiological parameters can be recorded with reference to heart and lung function.

VIII. Electromyography

Here, too, the treadmill is most useful. While the subject walks the activities of a number of muscles, or muscle groups, can be measured by means of an electromyograph, and recorded as a function of the temporal factors. This gives information about the function of the various muscle groups during the walking pattern. In other words, the only use here is to determine whether or not a muscle or muscle group acts during walking.

All the methods mentioned here are exclusively for the measuring of the movements of the lower extremities. The intention is to use analogous methods for the research of the upper extremities.

Authors' address: Dr. F. van Faassen, N. H. Molen, M. D., and W. Boon, Institute for Biomechanics and Rehabilitation, Free University of Amsterdam, *Amsterdam* (Netherlands)

Medicine and Sport, vol. 6: Biomechanics II, pp. 266–271
(Karger, Basel 1971)

Supporting Forces Related to Overload Damage of Leg Amputees

H. KRUSE, W. BAUMANN, and H. GROH

Institut für Biomechanik, Deutsche Sporthochschule, Köln-Müngersdorf

We tried to fix the form of load and its intensity on amputees by recording the rolling effects of feet and measuring their supporting forces. The experiment covered 70 test persons: 30 normal subjects, 20 shank amputees, and 20 thigh amputees.

Methods of Measuring

The supporting force of a foot may be regarded as a measure for the load exerted on the movement apparatus. In order to measure the supporting forces we developed a stationary force plate which renders possible the recording of the chronological process of the supporting forces in both the vertical and the horizontal direction.

Results of Measuring

The total of horizontal and vertical propulsing forces is indicated as support of the foot, the result being illustrated in figures 1, 2 and 3.

Supporting Force

In the evaluation, the force values A–E measured in the 70 subjects are related to the respective weights (=100%). Of those standard values we formed the mean $\bar{x}$, the standard deviation s, and the ratio of standard deviation to the mean

$$V = \frac{s}{\bar{x}}$$

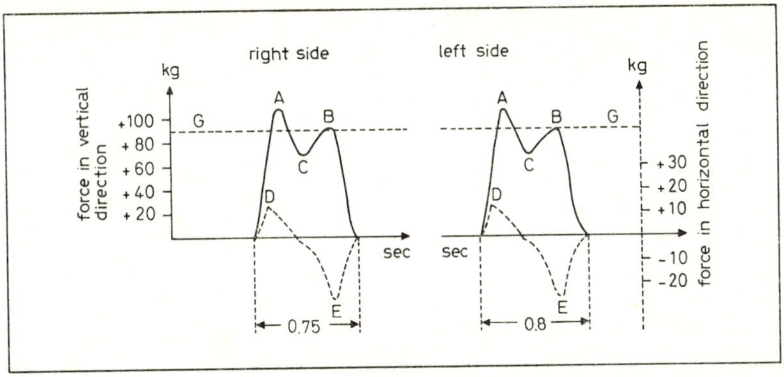

Fig. 1. Supporting forces, measured at the feet of a healthy subject.

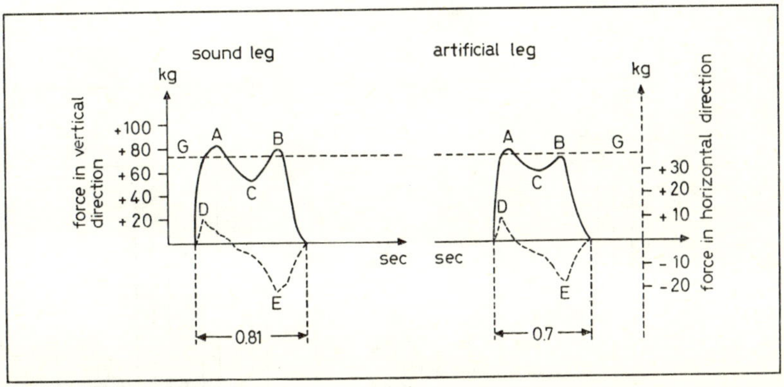

Fig. 2. Supporting forces, measured at the feet of a lower-leg amputee.

Figure 4 shows 5 maxima and minima A–E compared with the body weights (=100 %) of the 3 examined groups. The hatching indicates the scatter. Figure 5 shows the divergence of the 3 examined groups expressed as a percentage; force values of normal subjects being 100 %. The hatching marks the results of statistical significance.

Compared with healthy subjects, when the foot pushes off, the unloading of the sound leg as well as of the prosthesis is, on an average, 9 to 10 % smaller. This means a certain increase in the load.

The vertical pushing-off force B of the ball of the foot of the prosthesis leg is 6–7 % smaller than that in healthy subjects. The horizontal

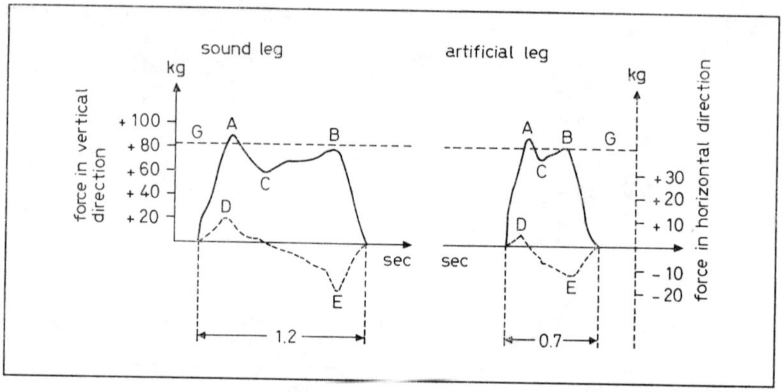

Fig. 3. Supporting forces, measured at the feet of an upper-leg amputee. Note: The left side of this figure represents the forces measured at the sound leg. The right side represents the forces measured at the artificial leg.

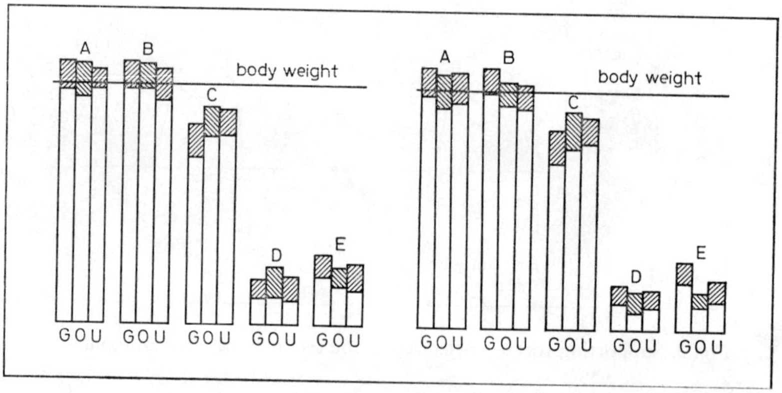

Fig. 4. The ratio between the actual supporting forces and the supporting forces due to a body weight of 100 kg. *G* Healthy, *O* upper-leg amputee, *U* lower-leg amputee.

pushing-off force in the walking direction in the case of the amputees is clearly smaller than in healthy people, viz. 20% less in the case of the sound leg, and 51 and 36% respectively in the prosthesis. The surplus load C of the amputee in the pushing-off phase is only of short duration and always stays below the maximum values (A, B) of normal subjects. The reduction of the pushing-off forces B and E clearly points to a smaller load in the gait of an amputee.

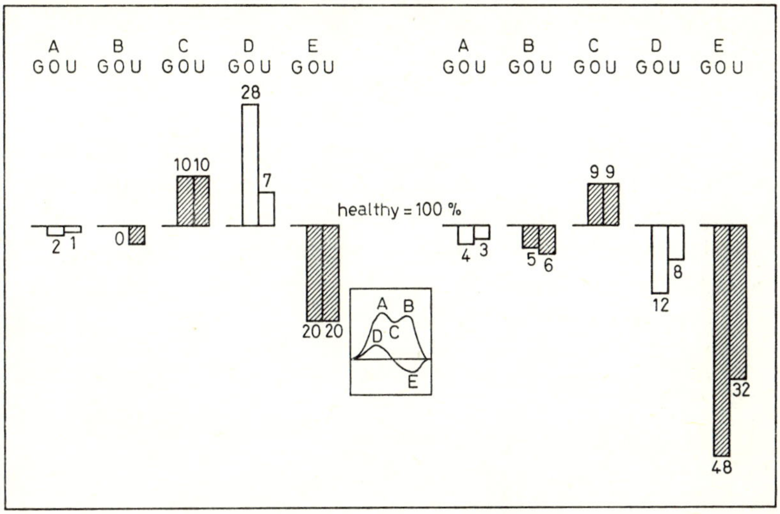

Fig. 5. Relative deviation of the supporting forces. Note: same as shown in figure 4.

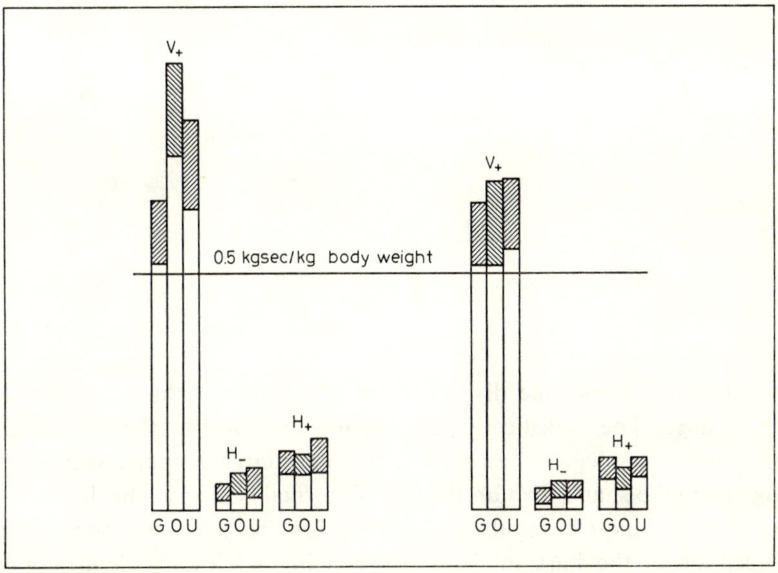

Fig. 6. Impulse of the force related to a body weight of 100 kg. Note: same as shown in figure 4.

Force Impulse

In addition to the effect of the maximum force, the duration of the load also plays an important part in overloading. For this reason we calculated the force impulse as force effect with respect to time for all force curves measured by integrating the force-time function by planimetry.

Figure 6 shows the different force impulses of the 3 examined groups: vertical impulse V+, retarding and accelerating horizontal impulses H– and H+. The hatching marks the scatter.

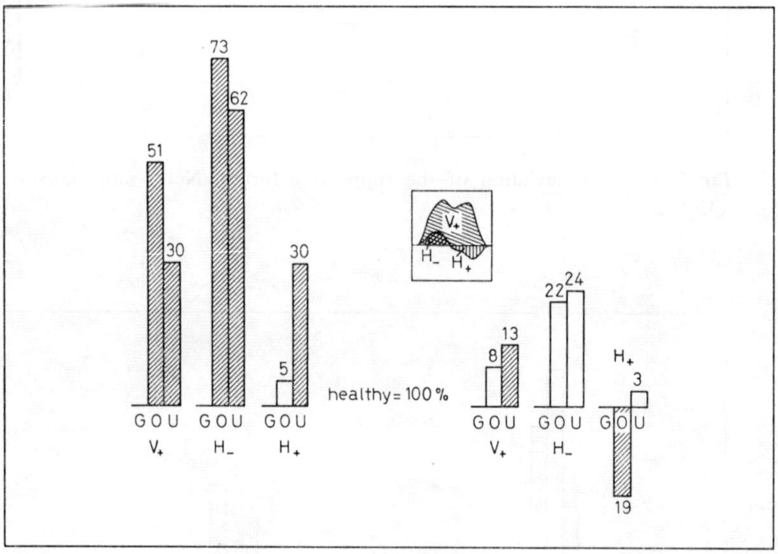

Fig. 7. Relative deviation of the impulse of the force.

Figure 7 shows the divergence of the impulse values expressed as a percentage. The statistically significant values are marked by hatching. On the healthy leg of the amputee there are larger impulse values on the thigh and the shank than on the legs of normal subjects. This holds for the vertical pushing-off impulses as well as for the horizontal ones, with the exception of the horizontal acceleration impulse of the thigh amputees. They are within the range of 24–64% above the corresponding values for normal subjects.

Asymmetry of Supporting Forces and Impulses

In the leg-amputees there exist asymmetry of horizontal supporting forces, vertical braking impulses, and of the amplitude of swaying of the body's centre of gravity.

Authors' address: Dr. H. KRUSE, W. BAUMANN, and Dr. H. GROH, Institut für Biomechanik, Deutsche Sporthochschule, Carl-Diem-Weg, *5 Köln-Müngersdorf* (FRG)

Medicine and Sport, vol. 6: Biomechanics II, pp. 272–276
(Karger, Basel 1971)

Bilateral and Unilateral Movement Patterns in a Mentally Retarded Child[1]

MARLENE ADRIAN and B. OGLESBY

Introduction

Locomotor patterns of mentally retarded children have been identified as inferior in quality to those of normal children [KEOUGH, 1969; FRANCIS, 1960]. In particular, a bilateral leg extension pattern such as the standing long jump is performed poorly. An immature asynchronous pattern tends to persist as though the bilateral action were too complex for the neuromuscular system of these retardates.

It is the purpose of this investigation to determine by means of electrogoniometry and electromyography the interplay of muscular contraction of agonists and antagonists during bilateral and unilateral knee extension patterns in a mentally retarded child.

Methods

A. Subjects

The subject was a 10-year-old trainable mentally retarded boy who possessed an immature standing long jump pattern. He weighed 47 kg and was 147 cm tall.

B. Instrumentation

Muscle action potentials and knee flexion and extension patterns were recorded on four channels of a Physiograph ink-writing oscillograph with a paper speed of 5 cm/sec. Electrogoniometers were used to measure flexion and extension at the knee joints. Each consisted of a potentiometer attached to a chassis and powered by a battery. The potentiometer was aligned with the axis of rotation of the knee joint, and the chassis was strapped to the body parts forming the joint.

1 Acknowledgement is made to B. CLEVELAND, B. FECHT, and M. MOWATT.

Surface electrodes of the pure silver disk sunken-well type from E & M Instruments, Inc. were used to measure the action potential of the rectus femoris and biceps femoris muscles. They were attached to the subject with disposable, double-sided adhesive washers according to standard procedures for bipolar electrodes, and placed 2 cm apart in a line parallel to the muscle fibres. The first muscle motor point was located by stimulation with a Medco-sonalator. The remaining motor points were located through palpation during manual muscle testing because the subject disliked being stimulated. No differences were found when a comparison was made between these two methods in the right rectus femoris. Electrode placement was identical for both right and left legs.

C. Testing

Although only two pairs of surface electrodes were used at one time, electrogoniometers measured flexion and extension of both legs throughout all testing. The movement patterns selected for testing were unilateral and bilateral knee flexion and extension on the LaBerne Exerciser and during stationary bicycling and jumping for distance.

Analytic procedures were determined by the limitations of the Physiograph System. Since the frequency response of the Physiograph was too low for valid quantitative assessment of the action potential, only sequence, amplitude, and duration of muscular contraction were identified and compared. A maximum contraction test was conducted to determine maximum amplitude of the electromyograms, so that comparisons in terms of percent of maximum could be made.

The goniograms provided data concerning range, angular velocity, and synchrony of movement. However, since the electrogoniometers were calibrated as 30 to 40 °/cm, there was a possible measurement error of ±5 °.

Results and Discussion

A. Synchrony of Bilateral Knee Extension and Flexion

The initial bilateral knee extension and flexion movements on the LaBerne Exerciser were asynchronous. The left leg extended and flexed prior to the right leg an average of 0.44 and 0.20 sec respectively. Furthermore, the ranges of motion were dissimilar.

When the knee was flexed and extended passively through its range of motion, the movement was synchronous. Both legs started in extension (180 °), flexed 100 ° to an angle of 80 °, and returned to extension. No evidence of muscular contraction was apparent.

Subsequent recordings were obtained for 12 bilateral extensions against resistance. In all but two instances the right and left movements were synchronized, but greater amplitude was noted in the left leg. The activity of the right leg terminated before complete extension while the left rectus femoris continued to contract.

The subject had difficulty performing bilateral flexion against resistance. He extended his knee and tried to push down on the resistance bar by extending the hip joint. Consequently, little contraction was noted in the biceps femoris. When the angle at the knee measured 155° or less, however, the flexion velocity increased and action potential was recorded.

B. Role of Agonists and Antagonists

Reciprocal innervation was present when unilateral and bilateral knee flexion and extension were performed against resistance (fig. 1). The rectus femoris contracted during extension and relaxed during flexion. At the exact instant of relaxation of the rectus femoris, the biceps femoris contracted; and conversely, the latter muscle relaxed when the rectus femoris contracted.

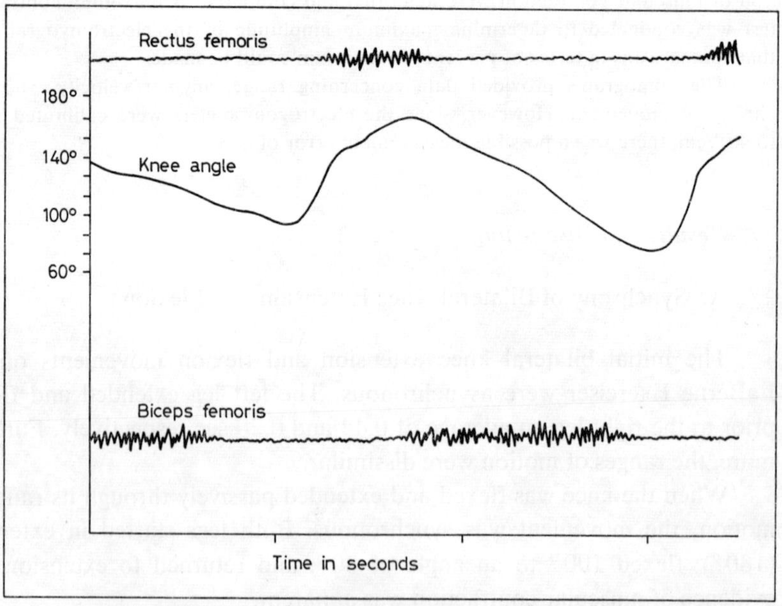

Fig. 1. Electromyograms and electrogoniograms of a mentally retarded boy's unilateral knee flexion and extension patterns on the LaBerne Exerciser.

This same phenomenon was noted when the subject rode a bicycle. In this instance, however, the contraction of the rectus femoris began an average of 14° before flexion was completed, and ended at the termination of extension.

C. Synchrony of Bilateral Extension during Jumping

In the standing long jump, there was lack of synchrony in all but one of the twelve jumping patterns. A typical recording is depicted in figure 2. Greater contraction of the right rectus femoris occurred and the left leg extended first with a greater range of motion. There was maximum contraction of the right rectus femoris but only 75% of maximum contraction of

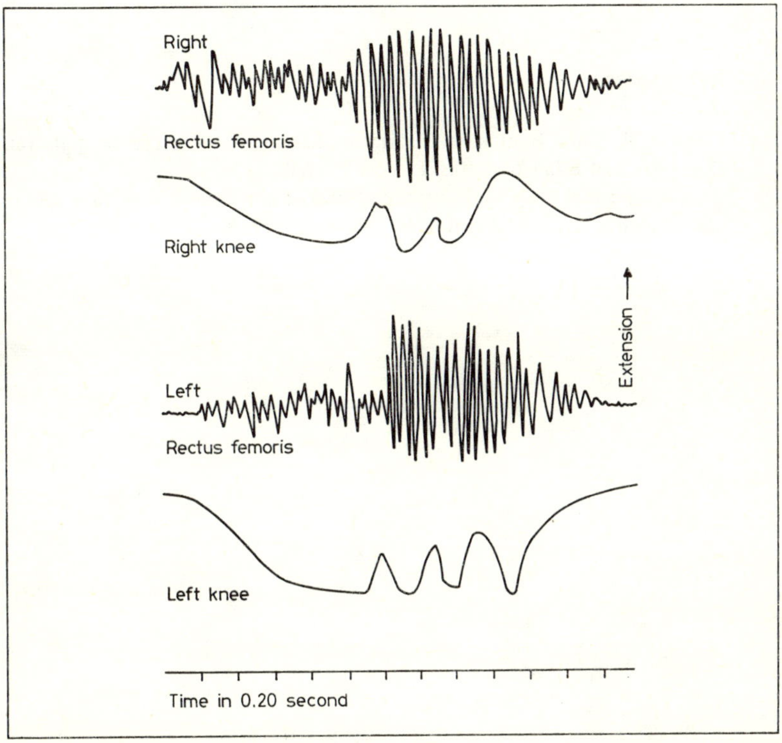

Fig. 2. Electromyograms and electrogoniograms of a mentally retarded boy's performance of a standing long jump.

the left rectus femoris. In all jumps the muscle contracted at the instant of extension. The left knee had an average amplitude of 62° whereas the right knee showed only 53°. The delay in right leg extension was 0.05 to 0.25 sec. All jumps were less than 8 in in distance.

Conclusions

Since the interplay of agonists and antagonists showed no abnormalities and a synchronous bilateral pattern occurred when the subject was in a sitting position, this suggests that no neurological impairment existed and that learning occurred. However, the standing long jump elicited an asynchronous response which may be caused by the lack of adequate leg strength to project the body when supporting the entire body weight or inability to control the position of the body center of gravity. The asynchrony may also result from the inability to utilize sequentially the various joints in a summation of forces.

References

1. FRANCIS, R. J. and RARICK, G. L.: Motor characteristics of the mentally retarded. U. S. Office of Ed., Coop. Res. Project 152 6432 (1960).
2. KEOUGH, J.: Incidence and severity of awkwardness among regular school boys and educationally subnormal boys. Res. quart. *39:* 806–808 (1969).

Authors' addresses: Dr. MARLENE ADRIAN, Department of Physical Education for Women, Washington State University, *Pullman, WA 99163;* Miss BILLIE OGLESBY, Glendora High School, *Glendora, CA 91740* (USA)

Medicine and Sport, vol. 6: Biomechanics II, pp. 277–284
(Karger, Basel 1971)

Programmed Stimulation of Skeletal Muscle – Some Problems

M. MILNER and A. O. QUANBURY

Control Systems Laboratory, National Research Council of Canada, Ottawa, Ontario

I. Introduction

Our work is essentially concerned with assessing the possibilities of, and developing, an electro-stimulation programmer unit for driving the musculature of the lower limbs of the paralysed to enable them to walk. We are currently establishing facilities [1, 2] for locomotion and allied studies which should facilitate the acquisition of program data for refined methods of feedback control.

Areas Explored

To illustrate our basic ideas about programmed electro-stimulation we have filmed results of experiments on direct stimulation of frogs' nerves in the pithed animal. This shows that controlled articulations of the leg and foot can be obtained with programmed sequences of stimulation at four sites which when individually stimulated produce gross actions from tibialis anterior, gastrocnemius, quadriceps and hip flexors.

We have investigated [1, 2, 3] the simultaneous effects of force and pain on normal subjects during electro-stimulation and have filmed various sequences of movements. These sequences are described later. It is considered that our studies thus far will serve as a frame of reference for further studies intended on incapacitated subjects in different categories.

Various problems can arise with electro-stimulation of a number of distinct muscle sites. Among the most salient are isolation of the several stimulus sources from each other; 'cross-stimulation' of antagonist muscle groups due to the field spread of current; a determination of whether there is effective 'motor-point stimulation' unaccompanied by reflex action, or purely reflex action or a combination of these effects. Stimulus isolators of various designs will eliminate the first-mentioned problem and we are investigating the other effects. Our approaches to some of these problems are outlined.

Being concerned with the control problem where feedback of information relevant to the state of a limb and its associated musculature is of importance, we have posed the question: 'Is it possible to transduce joint angle and force data from the electromyographic (EMG) responses produced by electrically stimulated muscle?' We make mention of some of our studies in attempts to elucidate this.

Finally, we render a brief outline of our proposed system for automating the derivation of detailed locomotion data from photographic records.

II. The Film

A. Frog Nerve-Muscle Stimulation

Individual and combinational stimulus trains were applied to nerves exposed in the pithed frog via small cuff electrodes made from short lengths of small bore plastic tubing with two fine insulated Karma wires threaded into the plastic. The insulation on each of the wires inside the cuff was carefully scraped off to effect electrical contact with the nerve. After all of the cuffs had been put in place the surgical wound was sewn closed and the animal supported by its forelegs so as to hang vertically with its back feet placed upon a horizontal rectangular board. The wire ends remote from the cuffs had their insulation removed and were wound around pins stuck into the board. The pin pairs were supplied via photon coupled stimulus isolators [1] which were driven from a 4-channel programmable stimulator [1] which we have constructed. The output characteristics of the stimulus pulses in the trains were arranged to be approximately 1.0 volt peak, 0.2 msec wide at 50 Hz.

Sequences of Actions

(i) Gastrocnemius, pushing off; (ii) lifting the leg, from the hip; (iii) tibialis anterior, lifting foot; (iv) tibialis anterior, lifting foot; (v) quadriceps, straighten leg; (vi) quadriceps, straighten leg; (vii) combined action of all four sites. The program employed was adjusted to produce the following sequences of actions: push-off on toes, lifting the leg, lifting the toes whilst lifting the leg, straightening the leg whilst being lifted, dropping the straightened leg, complete relaxation, repeat. Repetition rate was 0.5 cycle per sec (cps); (viii) speed-up of combined action. With the same program settings, the repetition rate was progressively increased until at about 10 cps, the combined effects of limb inertia and insufficient overlap of muscle responses to stimuli caused the limb to vibrate uselessly. The repetition rate was then re-adjusted to the original value; (ix) combined action.

B. Electro-Stimulation of Normal Human Subjects

1. Force and Pain

Using surface electrodes, we have thus far been able to effectively drive gastrocnemius, tibialis anterior and the quadriceps group in normal human subjects at tolerable pain levels. As an example of the character of the forces generated, it appears possible to cause the normal subject to be lifted up on his toes when left and right gastrocnemius muscles are stimulated simultaneously.

In our force/pain studies we have used electrodes of various areas (1, 2, 4, 8 in^2) and stimuli consisting of square wave voltage pulses of 50 Hz, 0.2 msec wide and going negative at the stimulus site. These pulses invariably were applied in trains lasting for 1 sec with an intervening rest period also of 1 sec. This regimen corresponds roughly to the periods of activity of the various muscle groups in an average walking cycle.

All the electrodes used in our studies, and in the film showing the human subject, were made from a fine stainless steel mesh whose interstices were filled with Redux electrode paste which was also rubbed into the skin at the sites of application. The cathodal electrodes were taped over the motor points of the selected leg muscles. The 12-in^2 anodal electrode was placed over the distal end of the stimulated muscle.

For gastrocnemius and tibialis anterior 11 normal male subjects were tested and in all cases the chief determinant of pain appeared to be the total peak current independent of electrode size. Peak currents in the range 40-105 mA were supplied and peak voltage levels fell in the range 15-40 V. Most subjects preferred stimulating electrodes at least 2-in^2 in area. Forces which can give rise to useful motions are elicited.

By using two active stimulating electrodes carefully placed experimentally over rectus femoris and vastus lateralis we have found it possible to consistently generate useful and substantial forces at the ankle with the straightening of the leg. In general it appeared necessary to maintain a small difference in potential between the two active electrodes. In this case 7 normal male subjects were tested.

2. Preparations for Filming

Various muscle groups were stimulated simultaneously in a female subject who volunteered. The electrodes described earlier were applied to the skin surface over the appropriate motor points after these had been determined using an exploring electrode having a saline soaked cotton pad surrounding a ¼ inch brass rod with an insulated handle. Stimulus drives were obtained via the isolators which in turn were driven by outputs from the multi-channel stimulator. Stimulus levels were adjusted to the just tolerable level and then reduced to the point where the motions resulting were positive.

The opening sequence shows the subject seated whilst the quadriceps groups of left and right legs are alternately stimulated for 1.5 sec at a time. Surrounded by the necessary equipment, the subject is seen kicking her legs in alternating fashion.

3. Further Sequences of Action (period 1.5 sec)

(i) seated, alternate stimulation of quadriceps group to produce kicking action; (ii) speed-up of the program in (i); (iii) standing with a chair as support with sequential stimulation of the quadriceps, tibialis anterior and gastrocnemius groups; (iv) speed-up of the program in (iii); (v) stimulation of 4 sites. Left gastrocnemius and right tibialis anterior together and right gastrocnemius and left tibialis anterior together. (vi) speed-up of (v); (viii) simultaneous stimulation of left and right gastrocnemius muscles, sufficient to lift the subject onto her toes. She uses a chair or the wall to maintain her equilibrium.

III. Some Problems

A. Surface Stimulation

To simplify the technique of electro-stimulation from the surface a prime requisite is for the muscle motor point to be close to the surface and isolated from sensitive organs. As mentioned earlier two active electrode sites are needed to effectively drive the quadriceps group. Thus careful field shaping appears necessary and this point also is of relevance if simultaneous stimulation of antagonist muscle groups is to be avoided. To gain some insight into the field patterns that can be generated, we have constructed an electrolytic tank model of the leg in the form of a clear plastic shell molded from the cast of the leg of a male adult. A molded model of the complete leg bone structure is mounted in the correct anatomical position and the tank filled with saline solution. Electrodes such as the type we are using can be held on to the inner surface of the plastic. We will report the results of our studies upon the model in detail elsewhere.

B. Transduction of Joint Angle and Force from Electromyographic Responses to Electro-Stimulation

It would be most desirable to transduce the state of a muscle and its allied joint(s) from the stimulating electrodes. However, it is first necessary to establish whether suitable transduction characteristics exist. To elucidate this point we performed an experiment [2] to study the relationships between volitional EMG, isometric force and angle at the knee joint when using the quadriceps group. Three sets of Beckman bipolar recording electrodes were attached to the quadriceps group; one set each approximately over the main bulk of the vastus lateralis, vastus medialis and rectus femoris. The EMG signals from the 3 sites were each amplified, summed and passed to a Hewlett-Packard Model 3400A rms voltmeter which outputs a d.c. analog of the meter reading. This output was applied to the Y-axis of an X-Y plotter, the X-axis being driven by a d.c. voltage proportional to the force produced at right angles to the ankle. Plots of summated rms EMG voltage against force were thus obtained for knee joint angles varying in steps of approximately 10° from 0 to 90°, and measured between the shank and the vertical. Fourteen subjects were tested and all rendered results similar to those shown in figure 1. Some relationship between force, angle and EMG voltage appears to exist but the mechanisms involved to produce the relationship have to be elucidated. We suspect that a considerable artifact is introduced by the relative motion of each muscle bulk under the electrodes on the skin. This might well be eliminated by using indwelling electrodes.

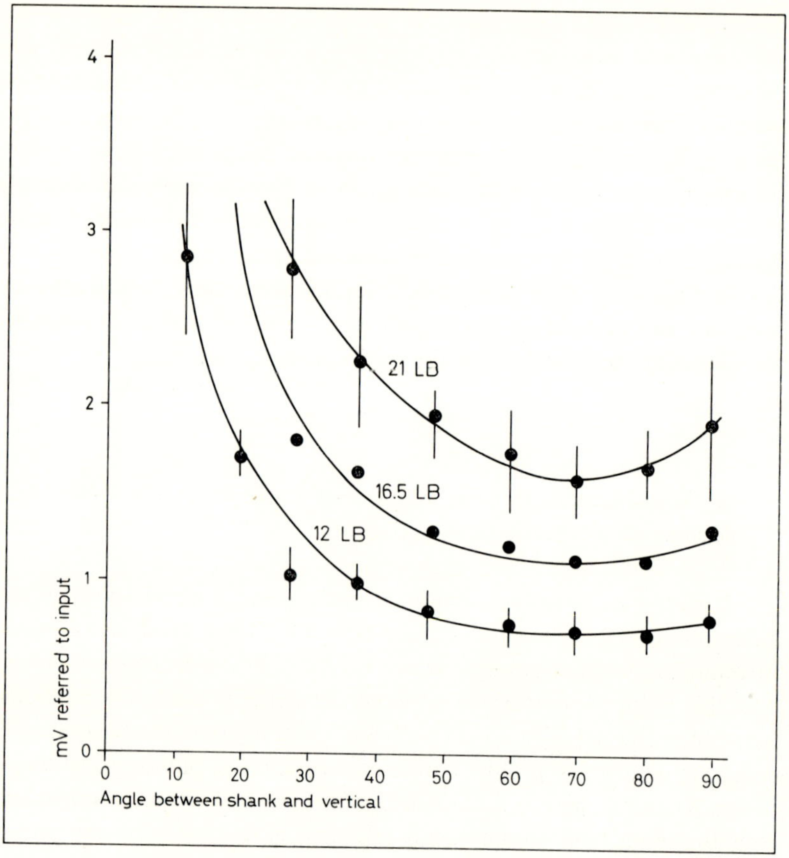

Fig. 1. EMG vs angle between shank and vertical for various constant forces at right angles to the ankle. Note: vertical lines show range of variation of EMG activity.

In making measurements of elicited EMG for individual pulses in trains of stimuli applied at 50 Hz with pulse duration of 0.2 msec we noted a general decrease in the EMG amplitude and a broadening of the waveforms as the angle between the shank and the vertical was increased. The artifact discussed above might be playing some role there too. Essential precautions to be taken when effecting measurements of elicited EMG are minimization of stimulus artifact by careful electrode placement and avoidance of amplifier blocking due to that artifact.

C. Derivation of Programming Data

We consider that the essential programming data can be derived from records of joint position trajectories and joint angles coupled with EMG activity. To facilitate the collection, processing and representation of the photographic data we are attempting to sequentially read individual 16-mm movie frames directly into our digital computer memory by the use of a flying spot scanner which we have available.

The subject will wear markers over various joints and will be photographed whilst passing in front of a background with a binary code of a suitable character [2]. The movie camera will move along with the subject. From the background each film frame will have distance information scaled and appropriate references with respect to the origin derivable, thereby enabling the markers upon the subject to be referred to the origin.

IV. Conclusions

While the completion of a programmer unit for electro-stimulation of the lower limbs to effect locomotion is some way off in the future it appears that several applications of the techniques of stimulation mentioned are possible. For example: programmed physiotherapeutic treatment where antagonist muscles may be arranged to act against each other and thus provide for loading of a weak muscle which it is desired to strengthen. Possible freedom of the physiotherapist who might supervise numerous subjects. Patterning of motions in various cases of brain damage. Provide for knee locking by electro-stimulation in cases of neurological disorders in which innervation or activation of the quadriceps is affected.

V. Acknowledgement

We are thankful to Dr. John H. Milsum who kindly presented our work at the Seminar.

VI. References

1 Milner, M.; Quanbury, A.O., and Edwards, E.P.: Progress Report No. 1. Human locomotion by ordered electrostimulation of the available musculature. Laboratory Technical Report, LTR-CS-6, Control Systems Laboratory, Division of Mechanical Engineering, National Research Council of Canada, April 1969.

2 MILNER, M. and QUANBURY, A. O.: Progress Report No. 2. Human locomotion by ordered electrostimulation of the available musculature. Laboratory Technical Report, LTR-CS-11, Control Systems Laboratory, Division of Mechanical Engineering, National Research Council of Canada, September 1969.

3 MILNER, M.; QUANBURY, A. O., and BASMAJIAN, J. V.: Force, pain and electrode size in the electrical stimulation of leg muscles. Nature, Lond. *223:* 645 (1969).

Authors' addresses: Dr. MORRIS MILNER, Department of Bioengineering and Medical Physics, Groote Schuur Hospital, Observatory, *Cape Town* (South Africa); Mr. A. O. QUANBURY, Control Systems Laboratory, National Research Council of Canada, *Ottawa 7, Ontario* (Canada)

Medicine and Sport, vol. 6: Biomechanics II, pp. 285–287
(Karger, Basel 1971)

A Muscle Stimulator for Hemiplegic Patients

J. VREDENBREGT and J. VAN LEEUWEN

Certain nervous disorders or damage to the brain can have the result that the muscles used in lifting the foot during walking are not activated, or are only partly activated. A patient with this disability (sometimes called drop-foot) is in danger of stumbling over his own foot, and when walking therefore tends to remain standing on the sound leg until the affected one is safely and securely on the ground. This gives him a dragging gait, and even with a walking-stick he moves only with difficulty.

It has been found that this disability can often be largely overcome if an electrical stimulus is applied at the appropriate instant to the nerve that causes the muscles involved to contract. A portable electronic muscle stimulator for this purpose has been developed and tested by the Institute for Perception Research in cooperation with the 'Stichting Revalidatie' (Centre for the Rehabilitation of the Physically Disabled), Eindhoven.

The main problem was to synchronize the periodically supplied stimuli with the patient's walk. In earlier investigations in this field experiments were carried out with a mechanical switch placed under the heel of the affected foot [1]. The disadvantage of this arrangement was that the stimuli were switched on and off at times which came too soon with respect to the desired cycle of lifting and lowering the foot. A better place for the switch was under the ball of the affected foot, but owing to the vulnerability and rapid wear of mechanical switches placed outside the shoe (and since these only work properly on hard level ground) the solution was not satisfactory.

We have now obtained very good results with a pneumatic switch which is operated by a small air chamber in a rubber insole fitted in the shoe on the unaffected foot. The air chamber is situated under the ball of the foot; the instants at which the ball of this foot touches and leaves the ground or, to be more exact, the instants at which the weight of the body is placed on the insole and removed again coincide very closely with the start and the end of the swing of the affected foot. The instants at which the muscles of this foot contract and relax are thus controlled correctly from the movement of the patient as he walks.

Figure 1 shows the appliance with the switch. The generator, i.e. the stimulator proper, is carried on a belt around the patient's waist, and delivers a rectangular pulse at a repetition rate of 50 Hz, 0.6 msec duration and a maximum amplitude of 100 V (rated current 6 mA). The amplitude of the pulse can be varied by means of a potentiometer. An elastic bandage immediately under the knee of the affected leg carries the two electrodes which are energized by the pulses. The positive (passive) electrode is made of electrically conducting rubber and is in direct contact with the skin. The other (active) electrode, to which a maximum of 100 V is applied, is made of stainless steel and makes contact with the skin through a damp sponge to avoid a high contact resistance.

The system has proved to be very reliable and the switch operates on any kind of floor covering or paved surface without getting damaged. This is a very important point, since it gives the patient confidence in the appliance.

For some considerable time the appliance has been tested at various centres for the rehabilitation of the physically disabled. As a result of using the appliance many patients have developed a more or less normal gait. An important contributory factor was found to be that the affected leg is better directed when the foot is placed on the ground and, therefore, stands more securely in the 'standing' phase. It is interesting to note that after some time many patients retain the improvement in their walk even without further use of the appliance. Out of 60 patients over 40 years of age who underwent the treatment for 3 to 6 weeks, 50 have recovered completely or show considerable improvement. The other patients responded well while walking with the appliance, but showed no lasting improvement. Two of these patients with no autonomous muscle control have been using the appliance constantly for more than 6 months, and are walking confidently.

The appliance is also being used on children with a drop-foot disability due to a brain defect contracted before or during birth. These children were not able to learn to walk normally. Some of these children who are paralysed on one side only (hemiplegia) have been under treatment for several months and have now begun to develop a more normal gait. For children paralysed on both sides (diplegia) an appliance with two separately adjustable outputs is now being tested; the insole here is placed under the less affected foot.

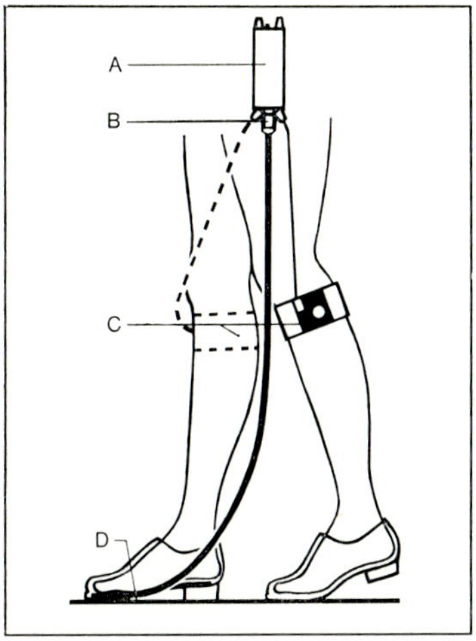

Fig. 1. Muscle stimulator. A = Generator, B = pneumatic switch, C = combined electrode, D = insole.

The results so far obtained with this appliance have demonstrated its helpful effect in the cases we have mentioned. It appears likely that the muscle stimulator will also be a useful aid for spastic children, and possibly for certain polio patients.

A semi-automatic appliance has now been developed at the Medical Systems Department of the Philips Concern. One of the advantages of this appliance is that the stimulation starts automatically when the patient begins to walk and stops when the patient stops walking. Moreover, the adjustment of the insole regarding the pressure in the pneumatic system is not critical, which contributes to a more convenient use.

Reference

1 Liberson, W. T.; Holmquest, H. J.; Scot, D., and Dow, M.: Functional electro therapy. Stimulation of the peronal nerve synchronized with the swing phase of the gait of hemiplegic patients. Arch. phys. Med. *42:* 101–105 (1961).

Authors' addresses: J. Vredenbregt, Instituut voor Perceptie Onderzoek, Insulindelaan 2, and J. van Leeuwen, Stichting Revalidatie Centrum Eindhoven, Kempensebaan, *Eindhoven* (Netherlands)

VI. Techniques and Methods

Medicine and Sport, vol. 6: Biomechanics II, pp. 290–298
(Karger, Basel 1971)

The Numerical Presentation of the Kinematics of Human Body Motions

W. GUTEWORT

Sports Science Section, Friedrich-Schiller-University of Jena, Jena

At the 1st International Seminar on Biomechanics we reported on the application of photogrammetry for recording human body motions [GUTEWORT, 1968]. The measurability of kinematic processes, i.e. the objectification of the morphology of human body motions, being important as a basis for biomechanical system analyses, the technique described has been advanced by an increase in measuring accuracy and an extended employment of electronic data processing.

1. Conditions of Stereophotographic Recording Using Pulsed Light

In the photogrammetric restitution of a photographically recorded model of a body motion, the *information content* is a function of the *space and time resolution* achieved. With n denoting the number of lines resolvable per mm, F the area utilized for the photograph and k the number of grey and colour tones distinguished at each point of the photograph, the spatial information obtainable is expressed by the relation $I_R = F \cdot n^2 \cdot \ln k$. Time information is given by

$$I_T = f_w \cdot \left[\frac{t_w}{t_B} \right]^{2/3}$$

with f_w standing for the frame or pulse frequency, t_w for the interval between exposures (pulse gap) and t_B for the exposure time per frame or pulse (pulse width). The product of space and time information might be regarded as the total information content [SCHARDIN, 1963]. From this aspect, classical cinematographic techniques are not likely to provide the most suitable solution to the problem.

The accurate *numerical presentation* of 3-dimensional paths and the determination of kinematic parameters and functions for rapid motions,

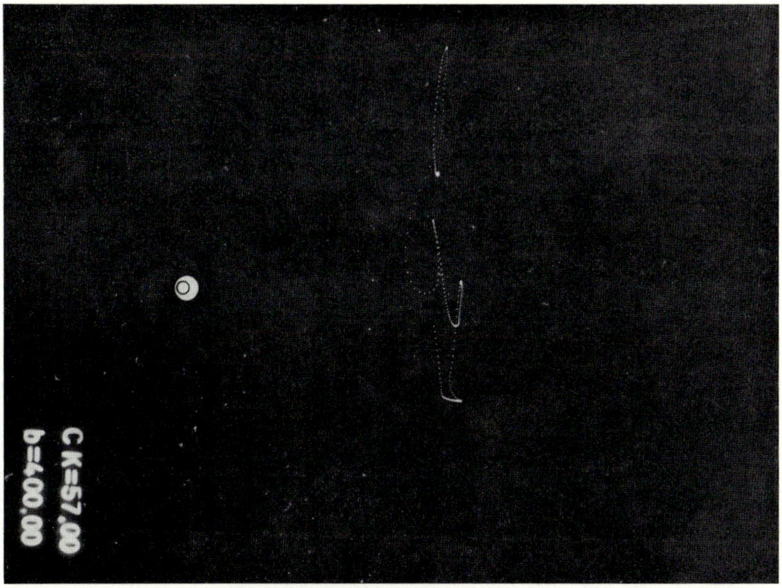

Fig. 1. Light pulse kinegram of the path of the bar bell in weight-lifting. Photograph taken with a stereometric camera, including photographed data on calibrated focal length and base length. The circular mark is one of the fiducial marks defining the centre of the photograph. The time interval Δt between the light signals was 5.16 msec.

whose space-time functions are unknown and whose parameters quickly change within short time intervals, are problems offering little access to measurement and calculation. At present, the only feasible way to implement the necessary resolution in space and time is by means of exact measurements made on a photographic model of the motion path.

The photographic recording of motions at a sufficiently high scanning rate and at properly defined measuring intervals included in the photographic record (fig. 1) was only possible by employing *photogrammetric stereocameras* together with highly stabilized *signal pulse transmitters* triggering electric lamps or electronic flash tubes that served to mark the measuring points [GUTEWORT, 1968, 1969].

The frequency fidelity and constancy of the electronic control systems thus achieved could not be obtained by means of mechanical clock generators. The timing error of the signal frequency is slightly above the accuracy of a quartz crystal (5–20 μsec, or approximately 0.05–0.1 %). The replace-

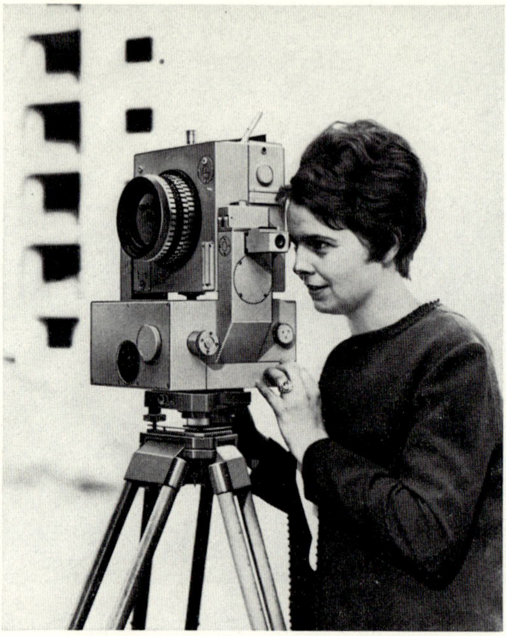

Fig. 2. UMK 10/1318 universal photogrammetric camera of VEB Carl Zeiss JENA. The Lamegon lens has a focal length of about 100 mm and can be set at infinity, 25, 12, 8, 6, 5, 4.2, and 3.6 m.

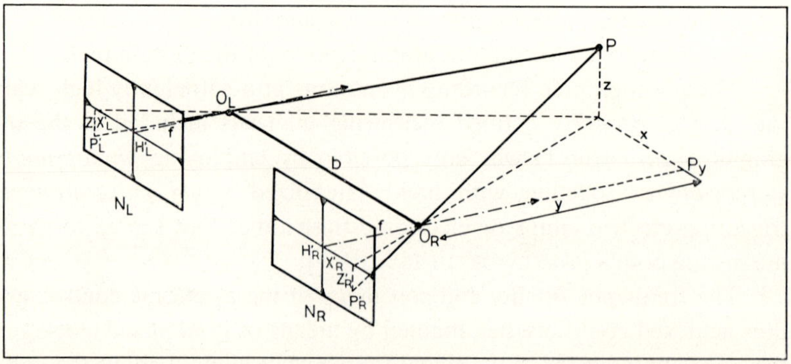

Fig. 3. Arrangement for stereocinegrammetry.

ment of square wave pulse shapes by spike pulse signals, emitted from the moving body in the form of light flashes, considerably reduced signal duration. Thus it was possible to avoid blurring due to motion, to increase resolution and, because of improved measuring conditions, to achieve conspicuously higher measuring accuracy.

The light signals were photographed with a UMK 10/1318 universal photogrammetric camera and an SMK 5.5/0808 stereometric camera of VEB Carl Zeiss Jena (fig. 2). Data compilation from the *kinematic photographs (pulsed-light kinegrams)* is effected by single-photo or stereo photogrammetry using comparators such as the Ascorecord or the Stecometer of VEB Carl Zeiss Jena. The coordinates of each measuring signal are determined and automatically recorded with a mean error of $\pm$ 1.6 to 2.5 μm. Notwithstanding the considerable increase in measuring accuracy, this technique allowed a substantial reduction of evaluation time as compared with the analysis of a cine film.

By comparing the coordinate parallaxes of the pairs of stereo-photo, all three coordinates can be calculated. From figure 3 it results that

$$Y = \frac{b}{p_x'} \cdot f,$$

$$X = \frac{Y}{f} \cdot x'_R = \frac{b}{p_x'} \cdot x'_R,$$

$$Z = \frac{Y}{f} \cdot z' = \frac{b}{p_x'} \cdot z'.$$

b = length of the camera base (distance between the optic axes)
f = calibrated focal length (distance between image plane and projection centre, identical for both cameras)
p_x' = horizontal parallax (difference between the abscissas of the images of point P in both negatives)
x'_R = horizontal difference between the image point P′ and the principal point H′
z' = vertical difference between the image point P′ and the principal point H′

2. *The Application of Electronic Data Processing for the Numerical Presentation of Human Body Motion Kinematics*

The Use of Computers in Digital Kinemetry

Given the space and time resolution attained in the photographic record, the application of electronic data processing in evaluating the photographs renders the coordinate values accessible to treatment by *differen-*

tial calculus. For that purpose, the function s = f (t) must be known as a prerequisite. The presumption proceeded from so far was that for the motions of the human body such functions cannot be found or are, at least, still unknown at present; so that only elementary mathematics in the form of step-by-step difference formation was employed [HOCHMUTH, 1967]. Given sufficiently high time resolution and measuring accuracy as well as constancy of the time interval (Δt), we are, however, able by making use of computing systems, to use this constant quantity as a parameter. In consecutive stages taking three successive points each time, the form of the curve can be approximated in parametric x(t) and y(t) presentation by a parabolic function through LAGRANGE interpolation.

With the function obtained it is possible by computation to determine further coordinate values of intermediate points lying between any two neighbouring measuring points and thus to reduce the measuring interval. The connection of these intermediate points by a traverse that closely approximates the actual curve makes possible the *calculation of arc lengths.*

Since the coordinates of each point except the starting and end values enter the calculation three times, smoothing and averaging of the empirically found values is effected simultaneously.

$$x(t)=a_1t^2+b_1t+c_1,$$
$$y(t)=a_2t^2+b_2t+c_2.$$

In an interval having a length of $t_0+2\Delta t$ (n=3), the resulting coefficients are expressed as follows:

$$a_i=\frac{H_{i_0}-2H_{i_1}+H_{i_2}}{2\Delta t^2}$$

$$H_{i_j}=\begin{matrix} x\ (t_0+j\Delta t) & i=1 \\ y\ (t_0+j\Delta t) & i=2 \\ & j=0,1,2 \end{matrix}$$

$$b_i=-t_0\,a_i+\alpha_i,$$

$$\alpha_i=\frac{-3H_{i_0}+4H_{i_1}-H_{i_2}}{2\Delta t}$$

$$c_i=-t_0\,b_i+H_{i_0}$$

By means of

$$dx/dt=2\,a_1\,t_0+b_1 \text{ and } dy/dt=2\,a_2\,t_0+b_2,$$

the *arc length* can exactly be calculated as the path

$$s_2\Delta t = \int_{t_0}^{t_0+2\Delta t} \sqrt{(2a_1t+b_1)^2+(2a_2+b_2)^2}\,dt$$

covered within the time interval $2\Delta t$.

Test calculations made on various curve configurations proved that in that manner the actual path is determined in very close approximation, making allowance for the path curvature.

Now the instantaneous rise

$$\frac{dy}{dx} = \frac{dx}{dt} : \frac{dy}{dt}$$

and the amount of the instantaneous speed at the time $t_0 + i\Delta t$ can be calculated from the equation

$$v_{i\Delta t} = \frac{ds(t)}{dt} = \sqrt{\frac{dx^2}{dt} + \frac{dy^2}{dt}}.$$

Data input into a digital computer is effected with the data being arranged on the punched tape as follows:

(1) Calibrated focal length (f) and base length (b) of the stereometric camera used for photographic recording. (2) Signal period Δt. (3) Coordinate values of the desired reference point to be constants. (4) As variables, the coordinates and coordinate parallaxes of each individual measuring point.

The following fully-programmed computing operations are possible: (1) conversion of Cartesian coordinates into polar coordinates and a defined reference point; (2) smoothing and averaging of the empirically found path-time functions and correction of controllable systematic errors; (3) determination of approximative functions for partial distances; (4) calculation of path elements, considering path curvature, which is an advancement over consideration of the traverse only; (5) calculation of instantaneous speeds and directions of movement by differential calculus, which is an advancement over step-by-step difference formation; (6) approximative calculation of acceleration and the instantaneous direction of its action; (7) separate calculation of specially relevant parameters; (8) calculation of the relative speed between individual joint axes; (9) calculation of the position of the body's point of gravitation and its path of movement; (10) calculation of individual body angles and the course of their change.

Further, electronic computation enables characteristic curves of mean value behaviours and, consequently, a *mathematical presentation of the motion patterns* to be abstracted from a number of measuring records of an identical motion element.

3. *The Measuring Accuracy of Pulsed-Light Photography*

The *error of speed measurement* must be determined from a practical example of a defined, precalculable motion, including the errors of all transmission links in the entire chain of information processing. Therefore, we have compared the calculated theoretical values of a defined motion (free fall, or pendulum swing) directly with the values measured empirically:

$$\delta v = \sqrt{\frac{[v_{i\,(theor)} - v_{i\,(emp)}]^2}{n}}$$

For cine-film pictures, the least possible mean error attained was

$\delta v = \pm 0.27$ m/sec (~3 %) at 24 frames/sec, and
$\delta v = \pm 0.44$ m/sec (~5.5 %) at 32 frames/sec.

For pulsed-light photographs taken with a 6×6-cm amateur camera and non-tested lenses, the error was

$\delta v = \pm 0.15$–0.05 m/sec (~1 %) without corrective calculation, and
$\delta v = \pm 0.088$–0.024 m/sec (~0.5 %) with corrective calculation.

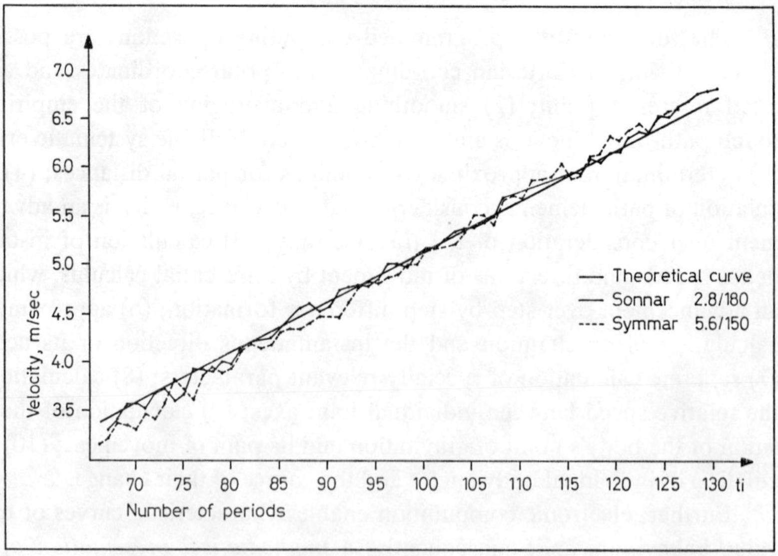

Fig. 4. Velocity-time diagram obtained with the aid of an amateur camera during the free descent of a weight of 1 kg. Length of the time period 5.240 msec.

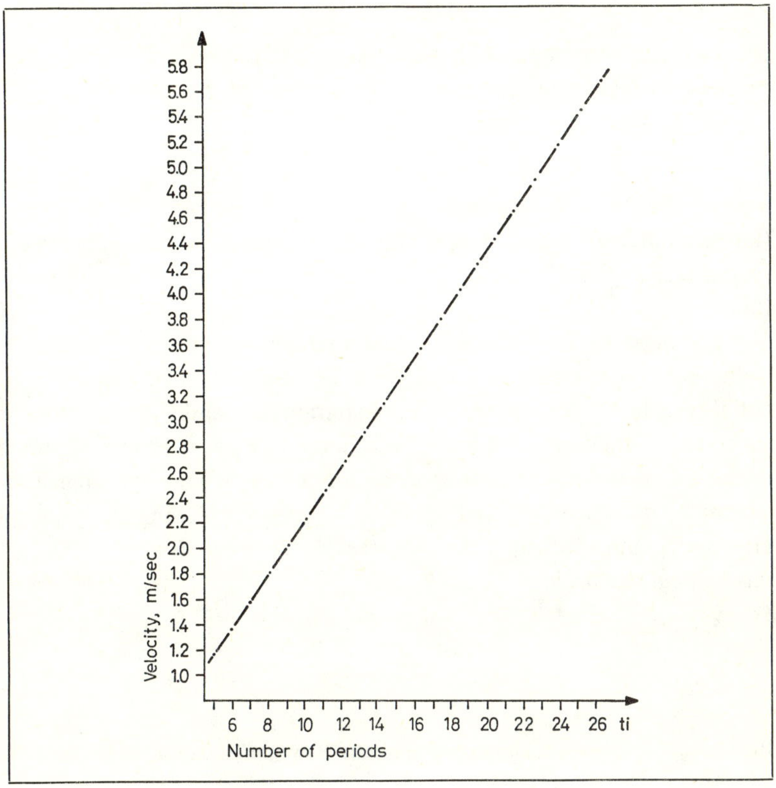

Fig. 5. Velocity-time curve obtained with the aid of the photogrammetric camera during the free descent of a weight of 1 kg. Length of time period 21.45 msec.

For kinematic photographs taken with a UMK 10/1318 Universal Photogrammetric Camera (photoplate format 13×18 cm), the error after corrective calculation was

$\delta v = \pm 0.0053\text{–}0.0072$ m/sec ($\sim 0.1\,\%$).

The difference in measuring accuracy between an amateur camera and a photogrammetric one is shown in figures 4 and 5.

The standard deviation of the *direction of movement* was calculated to be $\delta\alpha = \pm 0.09°$. BAUMANN [1968] found that the determination of *acceleration* from cine pictures is altogether unfeasible. This statement is in accordance with our own experiences made while using uncalibrated amateur cameras:

$\delta b = \pm 2.0\text{–}36.5\ \text{m/sec}^2$ (~20–350%!).

Using photogrammetric cameras and eliminating the errors by computation resulted in a decrease of the acceleration error down to

$\delta b = \pm 0.3\ \text{m/sec}^2$ (~3.0%),

so that pulsed-light photogrammetry appears to be suitable for *approximative acceleration determinations*. Thus, pulsed-light photogrammetry (kinegrammetry) should range among the most accurate kinematic measuring techniques known so far for motions having unknown space-time functions.

We hope to have made evident that, by solving the problem of the 3-dimensional measurement of human body motions, by employing differential calculus to determine kinematic parameters and by approximating the space-time functions of human body locomotions, it becomes possible to create *mathematical models of these locomotions*. Our experience may be summarized to the effect that *digital kinemetry,* employing analytical photogrammetry and electronic data processing develops a *new quality of biomechanical research.*

References

BAUMANN, W.: Über die kinematographische Bewegungsanalyse. Med. Welt, Stg. *40:* 2168–2174 (1968).

BEIER, W.: Biophysik. Eine Einführung in die physikalische Analyse biologischer Systeme; 3rd completely revised ed. (Leipzig 1968).

GUTEWORT, W.: Die digitale Erfassung kinematischer Parameter der menschlichen Bewegung. Biomechanics. I, pp. 53–60 (Karger, Basel 1968).

GUTEWORT, W. und TÖPFER, D.: Photographische Verfahren der biomechanischen Kinemetrie. I. Theorie und Praxis der Körperkultur, vol. 17, pp. 810–818 (1968).

GUTEWORT, W.: Fotograficzne metody kinemetrii w biomechanice. Sympozjum teorii techniki sportowej, Warsaw 1968, pp. 39–55 (AWF Sport i Turystyka, Warsaw 1970).

GUTEWORT, W.: Photographische Verfahren der biomechanischen Kinemetrie. II. Theorie und Praxis der Körperkultur, vol. 18, pp. 444–458 (1969).

HOCHMUTH, G.: Biomechanik sportlicher Bewegungen (Berlin 1967).

SCHARDIN, H.: Über die Grenzen der Hochfrequenzkinematographie. 5th int. Congr. High Speed Photogr., The Hague 1962, pp. 1–30 (Haarlem 1963).

Author's address: Dr. habil. W. GUTEWORT, Friedrich-Schiller-Universität, Sektion Sportwissenschaft, Seidelstrasse 20, *69 Jena* (GDR)

Medicine and Sport, vol. 6: Biomechanics II, pp. 299–303
(Karger, Basel 1971)

Human Motion: Simulation and Visualization

GLADYS GARRETT, W. S. REED, CAROL WIDULE, R. E. GARRETT

Purdue University, Lafayette, Ind.

Introduction

In the study of human movement, as in other areas of scientific research, there exists a recurring theme – a need for improved means of data quantification. With this explicit purpose in mind, research is being conducted jointly by the Departments of Physical Education for Women and Mechanical Engineering at Purdue University. The conceptual phase of the research is the result of the Women's Physical Education Department's efforts at focusing its attention on a program concerned with the preschool child's movement behavior. The need arose for some sort of a scheme that would enable the quantification of movement with enough precision to accurately assess individual differences in the movements.

In order to accurately evaluate differences in movements, where the data derived from cinematography is in a discrete form and is subject to the inherent errors associated with both time and position, it is necessary to work with a continuous mathematical model of the data rather than with bits of data. Because the Department of Mechanical Engineering is concerned with the mathematical modeling of physical systems ranging from gears and linkages to fluid and management systems, the extension of modeling to the human system provides both an interesting and challenging problem.

Procedures

The examples chosen for this paper are from a longitudinal study of a boy, David, filmed in slow motion at the ages of two, three, and four. A jump from a bench was selected because it possesses some good 'in flight' characteristics for checking the mathematical model.

With the exception of some body form data used in figure 2, the only data taken from the film were the joint coordinates at various selected frames. Various motion characteristics are generated by the computer and displayed graphically via an ink writing plotter or on a cathode ray tube. A movie was produced by refilming the graphical computer output of successive frames. In figure 1 is shown one isolated frame of this computer generated movie. The clock at the left, the stick figure, and the 'thermometer' charts at the right are self-explanatory. It should be pointed out, however, that since the movement is modeled, it is possible to obtain a 'frame' at any instant of time – not just those frames analyzed. Also, the selection of angular velocity shown in the charts was arbitrary. Any dynamic characteristic can be viewed in whatever form one desires.

The need for modeling arises from the nature of the data used to calculate velocity and acceleration characteristics of the body. Using the difference method to calculate velocity from data taken from film will often

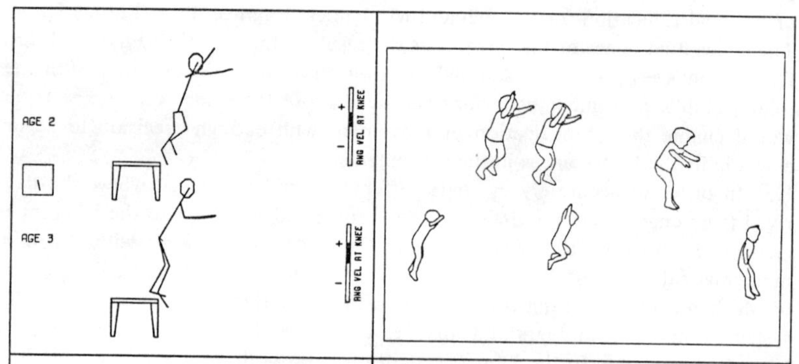

1, 2

3, 4

Fig. 1. Computer generated movie.
Fig. 2. Computer drawn profiles of jumping behaviour.
Fig. 3. Sequential movement of the head.
Fig. 4. Sequential movement of the lower leg.

5a 6a

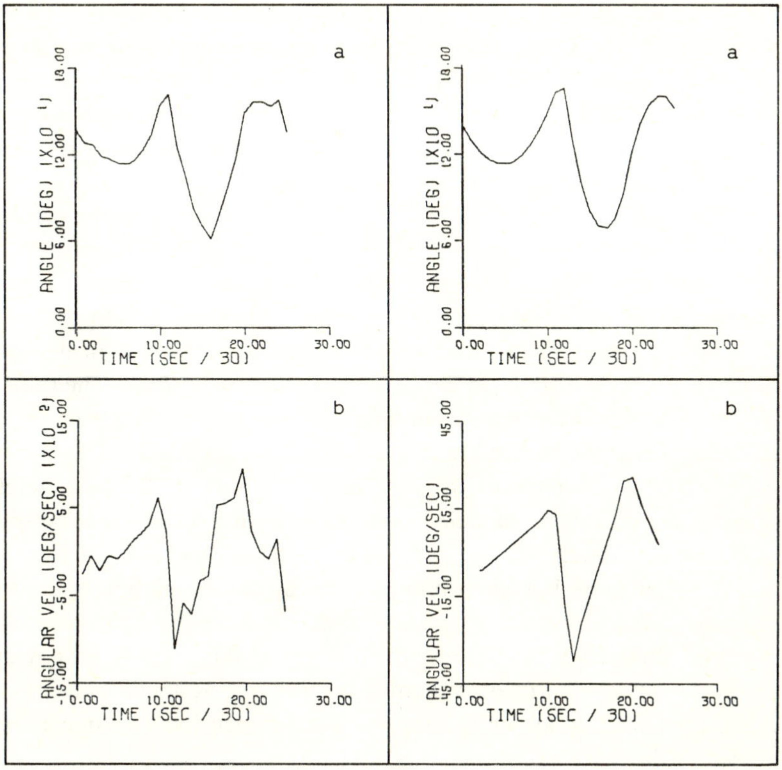

5b 6b

Fig. 5. Raw data.
Fig. 6. Modeled data.

produce large changes in velocity over short periods of time. This accounts for the rather ragged shape of the graph in figure 5b. If, however, the original raw data are replaced by a smooth curve closely approximating the original data, velocity and acceleration curves more consistant with theories of dynamics will result.

If an appropriate modeling procedure is used, a much better approximation to the velocity and acceleration will result. One such method is the second central difference technique derived from the Taylor series expansion. The modeled data and resulting velocity arrived at by using the second central difference technique are shown in the graphs of figure 6.

In figure 2 body forms digitized and stored on computer cards are shown. This type of graphical display is not only useful for recalling and comparing movement skills, but can also be useful for studying physical development. Although it is informative, this method of generating body forms is very time-consuming. The next logical step, therefore, is the complete computer simulation of human movement.

The computer simulation of the actual body throughout the prescribed motion is accomplished in three steps. The first step is to describe the solid body segments and a set of pivot points mathematically and to store this information in the computer. The next step is to read in joint position data and use the computer to generate the rotations and displacements needed to move the body from its initial position to its new position. The third step then is to perform the appropriate rotations and displacements on the solid body segments and to display this information graphically. The computer may then be used to predict the position of the body at all times, therefore, arriving at a movement modeled completely from a set of joint coordinates at various time intervals.

A rather gross but interesting attempt at this simulation is illustrated in figure 7. There are presently a few 'bugs' visible in this figure which have not been removed as yet because the concern at this stage is more with the efficient transmission of data than with error-free aesthetic drawings. The next step will be an attempt to represent a 3-dimensional body contour by a line drawing in two dimensions.

Fig. 7. Computer simulation of human movement.

At the present time both the CALCOMP Digital Incremental Plotter (an ink writing plotter) and the Control Data Corporation 252 Graphic Display Console (a cathode ray tube similar to a television screen) are used to display human motion characteristics in graphical form. The console is unique in that it not only provides an instantaneous graphic representation of any desired information, but also provides through an electronic light pen and keyboard a means for the true man-machine interaction so necessary in studying a complex system such as a human.

Authors' addresses: Dr. GLADYS E. GARRETT and Dr. CAROL J. WIDULE, Department of Physical Education for Women; Mr. WALTER S. REED and Dr. RICHARD E. GARRETT, Department of Mechanical Engineering, Purdue University, *Lafayette, IN 47907* (USA)

Medicine and Sport, vol. 6: Biomechanics II, pp. 304–307
(Karger, Basel 1971)

Quantitative Electromyography by Impulse Frequency Analysis (IFA)

H. U. DEBRUNNER

Introduction

A method for the measurement of the forces exerted on bone by skeletal muscles would be very useful in answering many questions which arise in biomechanics and in clinical practice. Such a method for the quantification of these actual forces does not yet exist. It is only possible to draw indirect conclusions from movement diagrams and force-time curves.

The electromyogram (EMG) records muscle activity. One can attempt to derive a relationship between the EMG and the forces developed by the muscle by integration of the EMG trace. Owing to the multiplicity of interference sources, however, it is not generally possible to draw reliable quantitative conclusions by this method. Integration of the EMG should, therefore, be regarded as a semi-quantitative method, depending on the measurement conditions.

Under particular confining conditions direct palpation of the tendon allows the estimation of the force in the tendon during a movement and thus to measure the degree of muscular contraction. This method, called myokinesography, was introduced into orthopaedic practice by SCHERB [1952]. It is technically difficult, however, and is, therefore, so limited in its application that it has found little acceptance.

Faced with another problem, we also became involved in the quantification of electromyography. The EMG is, for instance used for the control of electrical prostheses and appliances. In this case one is interested in a method of analysis of the voltage trace which is relatively insensitive to interference. If one wishes to achieve a proportional control a reliably reproducible electrical signal must be derived from the EMG. Integration is not very suitable for this purpose.

It is known that the count of EMG spike potentials during an isometric contraction is closely proportional to the force developed by the muscle [CLOSE, 1964, and others]. These spikes can be further processed as separate impulses, whereby the advantages which digital techniques possess over analogue methods can be used. In this way it would be possible to exclude a large part of the noise which is always

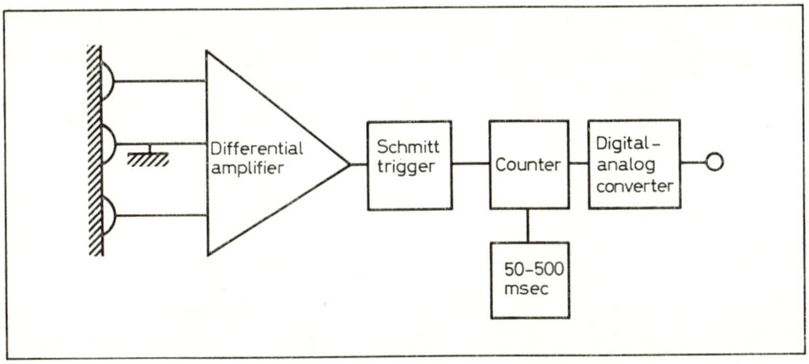

Fig. 1. Block diagram of method for impulse-frequency analysis of EMG.

present in an EMG signal. This led to a simpler and less interference-sensitive recording method for use outside the laboratory. Besides, the electrode techniques, which particularly bring problems where sensing is carried out over long periods (e.g. in prosthesis cases), have little effect on the result because the EMG is processed digitally.

Taking into account the considerations mentioned, we have evaluated muscle action potentials by a method which involves sensing and counting of the individual action potentials and conversion of the impulse sequence into a signal which can be recorded or can be used for proportional control of a prosthesis.

Method

The EMG is sensed using a gold-surface electrode (0.8 cm^2) and taken directly to a miniaturised differential amplifier with an amplification factor of 60 dB. This amplifier lies directly on the skin and its low output impedance allows the use of long connecting wires, and signal processing without feedback on the amplifier input. A variable-sensitivity Schmitt trigger responds to spikes which exceed the background noise level and after appropriate wave shaping passes them on to a small electronic counter. The time constant of this counter may be varied from 50 to 500 msec. A digital-analogue converter gives an output signal which is proportional to the frequency of the impulse train.

The use of the EMG for the control of electrical prostheses involves keeping the skin electrodes in the same place for several months, and in order to overcome this difficulty, we have recently successfully applied the Lopez-Richardson [1968] capacitative sensing method. In this way all procedures for preparation of the skin are done away with.

By using a pre-amplifier with a selective frequency response the sensitivity to interference, particularly mains-frequency hum, can be reduced. To wearers of prostheses working in industrial installations this elimination of interference is very important. In this way EMG investigations have thus become possible in working environments with strong electrical interference fields.

Results

The individual impulses which are above the noise level of the surface EMG have heights of 50–1,000 μV approximately, and durations of 5 to 10 msec. They are randomly distributed along the time axis. Their frequency is largely proportional to the degree of innervation of the muscle being investigated. When a rapid succession of impulses occurs, i.e. at high frequency, the impulses may coalesce to give a so-called interference pattern. For this reason the maximum frequency seldom rises above 300/sec. When innervation is minimal no spikes appear. A range can be found, characteristic of each muscle, within which the frequency of the train of impulses is proportional to the force exerted by the muscle.

Because the distribution of spikes with respect to time is stochastically independent, the counter time-constant should not be made too short. Counting for 100 msec gives good results: in this time 0, 20, 30 action potentials might be counted and 10 such countings would be possible per second. Thus extremely rapid movements, e.g. piano or violin playing, cannot be analysed but the succession of individual countings is sufficiently fast for the analysis of normal movements during walking and sport. When used for the control of prostheses the changes in impulse frequency with respect to time are followed so closely and the signal delay time is so short that good control is obtained.

The correlation between impulse frequency and muscle contraction force is sufficiently high. Under simple laboratory conditions we have obtained correlation coefficients of 0.8.

Comparison with the integrated EMG gives similar results. In strong interference fields near high power electrical apparatus the interference can be excluded better by the impulse-frequency analysis method by adjusting the threshold of the Schmitt trigger to the level of background noise. We then obtain a usable signal even under these difficult conditions. The same exists when the electrode set-up is not completely perfect.

Impulse frequency analysis allows standardisation of the output signal wherever the muscle contraction force can be directly measured. It probably has advantages in difficult circumstances.

References

CLOSE, J. R.: Motor function in the lower extremity (THOMAS, Springfield 1964).

LOPEZ, A. and RICHARDSON, PH. C.: Capacitive electrocardiographic and bioelectric electrodes. IEEE Trans. biomed. Engin. (1968).

SCHERB, R.: Kinetisch-diagnostische Analyse von Gehstörungen (Enke, Stuttgart 1952).

Author's address: Dr. med. HANS ULRICH DEBRUNNER, Abteilung für experimentelle Orthopädie Inselspital Bern (Switzerland)

Medicine and Sport, vol. 6: Biomechanics II, pp. 308–315
(Karger, Basel 1971)

Muscular Tonus Measurements

W. D. van der Laarse and W. J. Oosterveld

Vestibular Department, ENT Clinic, Wilhelmina-Gasthuis, Amsterdam

I Muscular Hardness Measurements

Introduction

The vestibular system has a definite effect on muscular tonus and hardness. In order to measure muscular hardness it was necessary to develop two instruments which would provide a definite evaluation of this hardness.

Muscular hardness is a general description which eludes clear definition. It denotes the hardness of a group of muscles or an individual muscle and depends on a number of factors of very disparate character such as the thickness of the skin, the blood pressure and the tonus of the myofibrillae at rest.

This rest tonus has aroused the interest of many physiologists over the years. It is difficult to measure muscular hardness and tonus. Several measuring techniques have been developed, but most of them had the disadvantage of yielding data which were not reproducible.

Wertheim Salomonson [1911] evolved a self-recording myosclerometer based on determination of the distance over which a pin impresses a muscle when a pressure of 2.75 kg was applied to the pin. The muscle tonus thus measured was expressed as a so-called hardness figure. Wertheim Salomonson applied this procedure in a comparative study of the hardness figures of the left and right calves in patients suffering from tabes dorsalis, hemipareses, muscular dystrophy, scleroderma, and multiple sclerosis.

Noyons and van Uexuell [1911] and Polée [1918] made their investigations with the aid of this technique.

Technique

On the basis of the principles of the Wertheim Salomonson technique an instrument was developed which makes the quantitative evaluation of muscular hardness a comparatively simple task (fig. 1).

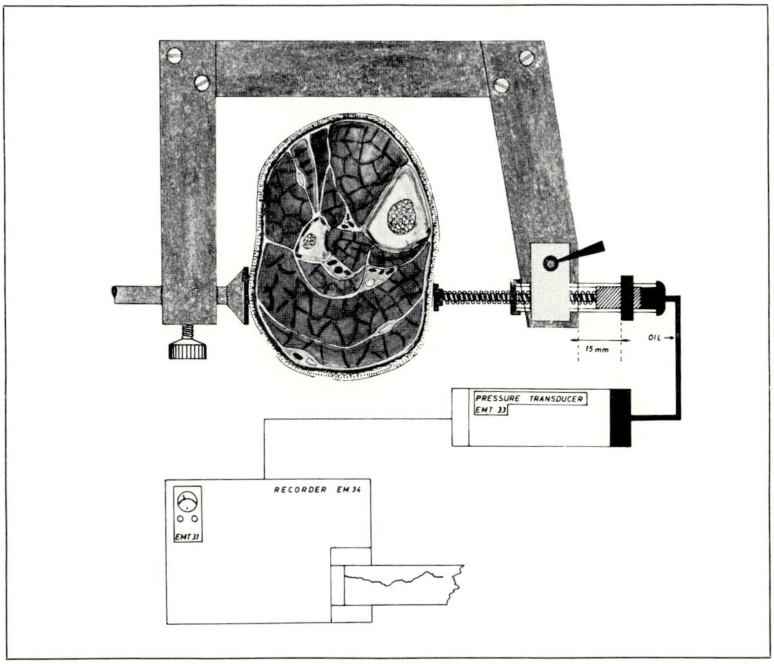

Fig. 1. Equipment for muscular tonus measurement (myosclerometer)

The device works as follows: A horseshoe-shaped brace is placed over the broadest part of the calf muscles of one leg. One arm of the brace shoe has an adjustable die whose correct position is determined and fixed for each individual subject. The other arm consists of a modified tuberculin syringe placed with the handle of the plunger facing the adjustable die and in line with the die axis. This syringe is fixed to one arm of the brace. Changes in the leg diameter are made with the adjustable die on the opposite arm of the brace. The modification of the syringe consists in the presence of two compression springs around the piston axis, one inside the glass cylinder and one outside; these ensure that the piston resumes its initial position immediately after the syringe has been used. The syringe also has a stop-cleat by means of which it can be displaced over 15 mm in the direction of the calf muscle in all experiments.

Method

The patient or the test person is laid on a stretcher, the knee and the heel are supported in such a way that the calf muscles are relaxed. Now the brace is placed over the broadest part of the calf. The syringe and the adjustable die both are in their extreme positions. Then the die is adjusted in such a way that it just touches the skin. The syringe is moved towards the skin over 15 mm and fixed with a screw. Caused by the muscle hardness a small proportion of fine oil (without air bubbles) is displaced towards the Elema pressure transducer EMT 33, and the increase in pressure is converted into an electrical potential which causes a proportional deflection of the manometer needle. The pressure can be read and recorded immediately.

In this way the hardness of the calf muscles was recorded in 12 test subjects (table I). Repeated determinations at the same site of the lower leg showed that the values measured were reproducible within narrow limits. The subjects showed comparatively large differences between the individual values.

Table I. Determination of hardness figures in 12 normal test subjects. The intra-individual range proved to be small

Subjects and ages												
Time of control	I	II	III	IV	V	VI	VII	VIII	IX	X	XI	XII
min	21	29	20	26	36	32	20	24	26	21	26	23
5	110	98	104	118	88	80	180	150	70	80	90	101
10	108	100	104	114	86	80	180	150	70	78	88	102
15	112	96	104	108	90	82	180	148	68	82	90	101
20	112	96	100	112	88	82	180	150	68	76	89	100
25	112	96	106	108	88	80	170	145	68	76	92	98
30	112	96	108	110	88	84	170	149	68	76	90	100
35	114	94	108	108	86	82	170	147	72	78	92	101
40	110	94	110	109	88	84	165	150	74	76	91	100
45	114	94	110	108	84	82	164	149	76	78	90	102
50	110	100	110	102	84	84	165	148	70	76	92	101
Mean	111.4	96.4	106.4	109.7	87.0	82.0	172.4	148.6	70.4	77.6	90.4	100.6

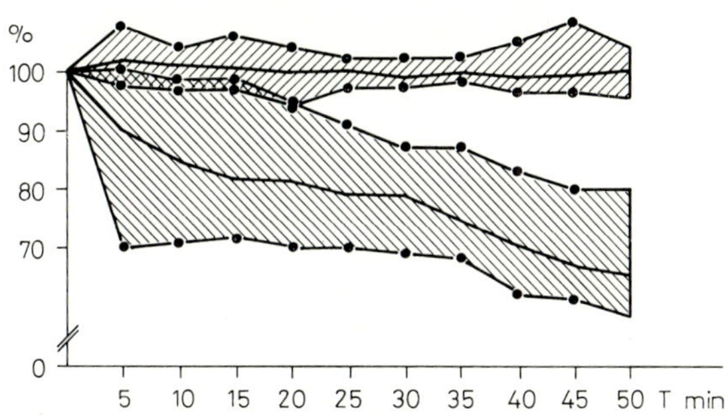

Fig. 2. The influence of 20 mg Valium on muscular tonus in 12 subjects.

In an effort to obtain some information on the possibility of influencing muscular hardness with the aid of a drug, the same 12 subjects were given 20 mg Valium (fig. 2). This drug has no influence on the blood pressure. The diminution of muscular hardness after Valium medication can probably be ascribed to its effect on muscular tonus. The instrument described enables reproducible information on muscular hardness to be obtained.

II Determination of the Degree of Spasticity

Introduction

The second equipment enables further information to be obtained by determining the degree of spasticity. Spasticity of the muscles in patients with a transverse lesion is a troublesome symptom. In order to combat it pharmaceutical, physiotherapeutic and surgical methods are applied.

To determine the degree of spasticity one can use the pendulum test described by MUMENTHALER [1965]. Studying the muscle-relaxant effect of diazepam (=Valium) in spastic patients, SCHLÄPFER and MUMENTHALER [1964] used the pendulum test which is as follows (fig. 3):

A light bulb is attached to the foot of the freely hanging and relaxed leg. The leg of the sitting patient is extended straight forward and then released. The lower leg with the attached light bulb follows a damped pendulum movement which is filmed from the side. This method has a few disadvantages:

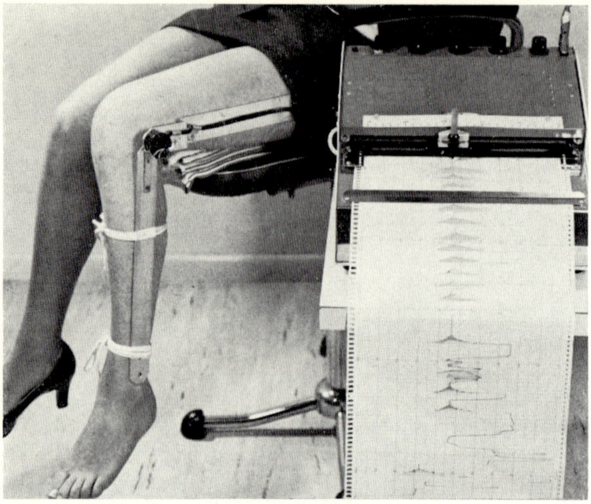

Fig. 3. The pendulum test equipment.

(1) The camera must be pointed perpendicularly to the plane of the arc described by the light bulb and the transport of the film must be in vertical direction and requires a calculated speed. If this requirement is not fulfilled, the curves may show some artefacts.

(2) The film transport must be continuous and exposure must be uninterrupted by a shutter. This implies the use of an expensive camera.

Technique

To simplify this pendulum test for clinical application and research purposes it was suggested that (1) the test should be simple and rapidly performable; (2) a darkened room should not be necessary; (3) the results of the test must be immediately available; (4) the curves should not need further processing before filing with the patient's records; (5) loops in the curves are not allowed; (6) the cost should be less than that of filmed curves.

Method

We were able to meet these requirements with the following apparatus. For recording we used a HEATH recorder, type EUW-20A. Any other available EEG or ECG recorder may also be employed. The paper speed in our experiments was 50 mm/min.

The movement of the swinging lower leg is transmitted to a simple motion transducer applied laterally to the knee joint. This transducer consists of a wire-wound potentiometer of 5 kΩ and two metal strips. One strip is fixed to the case of the potentiometer, the other to its spindle. Both strips are attached to the upper and lower leg with sticking-plaster or rubber bands in such a way that the axis of the spindle of the potentiometer coincides with the axis of the knee joint. The motion transducer is connected via a 1.5-volt battery to the recorder. Each movement of the lower leg leads to a variation in resistance which causes a proportional alteration in the current at the recorder input. This apparatus permits a rapid pendulum test; darkening of the room is not necessary; the curve is immediately available and can be filed in the patient's records. In this way loops cannot occur. This modification of the MUMENTHALER pendulum test enables quantitatively reproducible values to be collected suitable for further mathematical analysis. The values are obtained in the course of the examination in the form of curves drawn on paper and they can be directly evaluated by the examiner (fig. 4). Effects, if any, of drugs acting upon muscular spasm can be studied during the examination itself. As was done

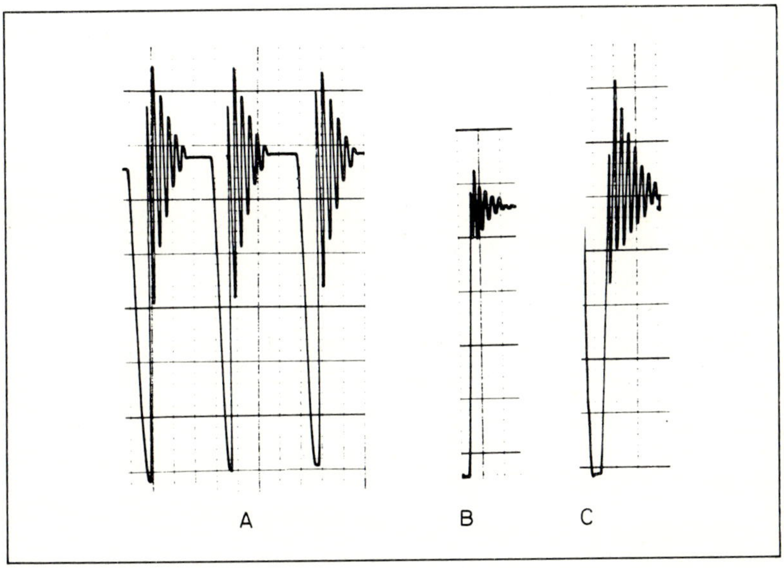

Fig. 4. A. Normal pendulogram; *B.* before, and *C.* after administration of Valium, in the same patient.

Table II. Normal persons (A) and patients before and after administration of Diazepam (B)

A

Test subject	Total pendulum time, sec	Number of swings	Amplitude
1	5.4	6	126
2	5.4	5	100
3	4.8	6	112
4	5.4	6	124
5	6.0	6	118
Average	5.4	5.8	116

B

	Before administration of Diazepam				After administration of Diazepam			
	Total pendulum time	Number of	Amplitude		Total pendulum time	Number of	Amplitude	
Patient	sec	swings	Measured	Corrected	sec	swings	Measured	Corrected
1	4.2	5	21	100	5.4	6	65	309
2	4.2	4	145	100	6.0	6	153	107
3	6.0	7	82	100	9.0	9	343	418
4	0	0	0	—	3.6	4	84	—
5	2.4	4	14	100	7.8	8	65	455
6	4.2	2	2	100	4.2	5	27	1350

with the myosclerometer discussed in part I of this paper, the data obtained from 5 normal subjects and 6 patients from the Rehabilitation Centre 'De Hoogstraat' at Leersum, Netherlands, were compared [COUVÉE, VAN DER LAARSE, and OOSTERVELD, 1968].

It is shown in table II that Valium has also an effect on the swinging movement of the lower leg, as was to be expected from the results mentioned in part I.

Conclusion

Both apparatus, the myosclerometer and the pendulum test device, have been developed in the Vestibular Department, ENT-Clinic, Wilhelmina Gasthuis, Amsterdam; they may be important for clinical investigations in patients suffering from spasticity. The method may also be helpful to the pharmaceutical industry for research in this special field of drug investigation.

References

COUVÉE, L. M.; VAN DER LAARSE, W. D., and OOSTERVELD, W. J.: Paraplegia *6:* 2 (1968).

MUMENTHALER, M.: 7. int. Kongr. Neurol., Wien (1965).

NOYONS and VAN UEXKUELL: Z. Biol. B. 56; Onderzoekingen Physiol. Lab. der Utrechtse Hoge School, 5e reeks, vol. 12 (1911).

ZWAARDEMAKER: Leerboek Physiol., vol. 2 (1911); description of WERTHEIM SALOMONSON's myosclerometer.

Authors' address: W. D. VAN DER LAARSE and W. J. OOSTERVELD, M. D., Physiological Laboratory, ENT Clinic, Wilhelmina-Gasthuis, Amsterdam (Netherlands)

VII. Miscellaneous

Medicine and Sport, vol. 6: Biomechanics II, pp. 318–322
(Karger, Basel 1971)

Kinematics and Tactics of the Hand Motor Preparation in a Serial Choice Task

MARIE-CLAUDE HENNEMANN

Centre d'Etudes bioclimatiques, Strasbourg

The present study concerns the motor preparation in a serial 5-choice reaction task. The subject's performance is recorded during two equal periods of 20 min. We have tried to decompose the time interval between two successive stimuli in order to analyse the basic behaviour involved in such a task.

The subject sits alone in a sound-proof room, in the center of a semi-circular display formed by 7 panels. On the central panel, in front of the subject, a Nixie Tube displays the digits 1, 2, 3, 4 or 5, appearing at the rate of 42 digits per minute. The digits are presented equiprobably in a random order, without successive repetition.

According to the displayed digit the subject must depress with his right hand the appropriate key of a keyboard without looking at it. The keys are located in numerical order on the keyboard. To depress the keys the subject is asked to use only his fore- and middlefinger and to respond as fast as possible without making errors. Pressing a key cancels out immediately the digit which is automatically replaced by another one.

The movements of the two fingers as well as a digital clock exhibiting time at 0.01 sec are recorded photographically by a camera set at a speed of 32 frames per second. Subsequently the films were analysed frame by frame. For each interval between two successive stimuli, several spatial and temporal data have been considered, making it possible to distinguish between the effective response and the preparatory period (fig. 1).

(a) The effective response period corresponds to the time elapsed between the onset of stimulus N and its extinction. This time can be split into (1) a latency time TL_R between the onset of stimulus N and the start of the hand movement towards the chosen key, (2) a movement time TM_R measured between the beginning and the end of the hand movement, (3) an

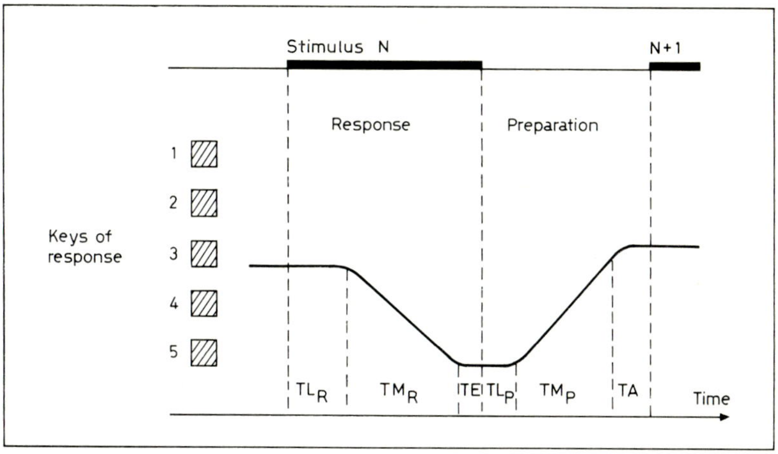

Fig. 1. Elementary times during the interval S_N, S_{N+1}.

effector time TE required to depress the key and measured between the end of the hand movement and the extinction of the digit.

(b) The preparatory period corresponds to the time measured between the extinction of stimulus N and the onset of stimulus N+1. This time can also be split into (1) a latency time TL_P measured as the time elapsed between the extinction of the digit and the beginning of a hand movement to reach another position above the keyboard. This movement is not required from the subject but proceeds from his own impulse; (2) a movement time TM_P measured between the beginning and the end of this hand movement; (3) a waiting time TA until the onset of the next stimulus N+1.

It appears that after every response and for all Ss, the preparatory movement consists in moving the hand towards the central position of the keyboard (fig. 2).

This pattern of motor preparation turns out to be a behaviour built-up during practice periods of the 5-choice task. Indeed, the films recorded during the first period of practice show that after depressing the adequate key, the subject keeps his hand over it, till the apparition of the next digit.

During the second period of practice, a strategy for preparing the next response builds up. Actually after each response the subject moves his hand back over the central position of the keyboard, approximately over key No. 3. The tactics thus adopted seem to be a consequence of the special

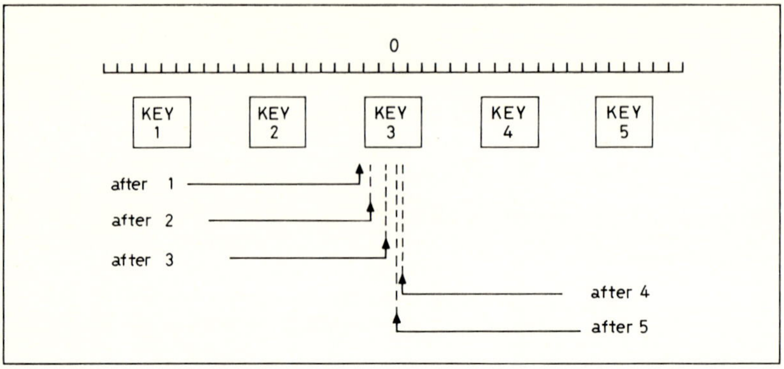

Fig. 2. Average position of the hand after responses to digits 1, 2, 3, 4 or 5.

instructions to work fast, given to the subject at the beginning of the second practice period.

Meaning of the Motor Preparation

The observed motor preparation shows that the subject tries to reduce the uncertainty of the next stimulus by choosing an appropriate starting position for the next response movement.

The keys being equidistant when referred to the wrist around which the hand turns, the length travelled by the hand is most easily expressed in arbitrary units, i.e. the angular distance between two adjacent keys.

In this way, two different tactics of preparation have been evaluated: the first when the subject's hand remains over the key he had previously depressed (tactics 1); the second when his hand returns to a central position of the keyboard (tactics 2) (table I).

In each case there are two algebraical values (the sign + means a movement to the right of the keyboard, the sign — means to the left). The first value in brackets is the distance described by the preparatory movement, the second is the distance described by the actual response movement. In the case of tactics 1: the sum of possible preparatory movements is zero because the hand remains above the depressed key; the sum of possible movements to respond to the digits which are presented is 40 arbitrary units. In the case of tactics 2: the sum of possible preparatory movements (back to central position) is 24 units; the sum of the movements of the hand to

Table I. Movements of the hand. Displacements in arbitrary units.

		Tactics 1						Tactics 2					
		Response n						Response n					
		1	2	3	4	5		1	2	3	4	5	
Response (n–1)	1		(0)	(0)	(0)	(0)	(0)		(+2)	(+2)	(+2)	(+2)	(8)
			+1	+2	+3	+4	10		–1	0	+1	+2	4
	2	(0)		(0)	(0)	(0)	(0)	(+1)		(+1)	(+1)	(+1)	(4)
		–1		+1	+2	+3	7	–2		0	+1	+2	5
	3	(0)	(0)		(0)	(0)	(0)	(0)	(0)		(0)	(0)	(0)
		–2	–1		+1	+2	6	–2	–1		+1	+2	6
	4	(0)	(0)	(0)		(0)	(0)	(–1)	(–1)	(–1)		(–1)	(4)
		–3	–2	–1		+1	7	–2	–1	0		+2	5
	5	(0)	(0)	(0)	(0)		(0)	(–2)	(–2)	(–2)	(–2)		(8)
		–4	–3	–2	–1		10	–2	–1	0	+1		4
Total movement													
before signal N		(0)	(0)	(0)	(0)	(0)	(0)	(4)	(5)	(6)	(5)	(4)	(24)
after signal N		10	7	6	7	10	40	8	4	0	4	8	24

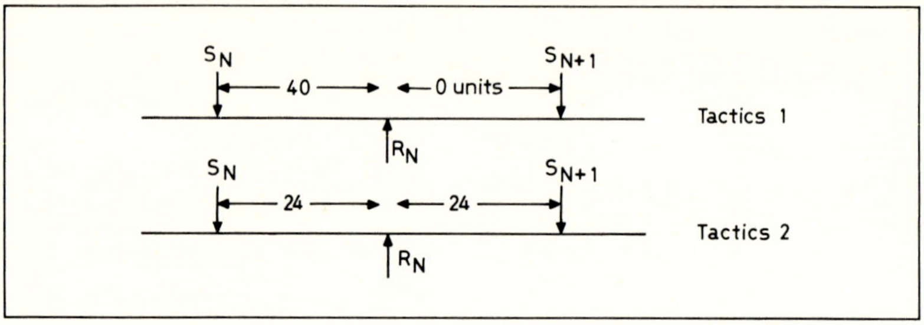

Fig. 3. Sum of possible movements of the hand according to tactics 1 and tactics 2.

respond from this central position is also 24 units. In this latter case, we have a total of 48 units instead of 40 units for the first tactics. Thus the subjects seem to prefer the more costly of the two tactics. However, it appears that this strategy is indeed better for speed because half of the displacement (24) is made before the onset of the next stimulus, leaving only 24 units of displacement to be travelled after the onset of the next digit (fig. 3).

This reduction in the total distance of actual response movements involves necessarily a decrease of the mean response time. Hence the tactics exhibited by the motor preparation complies with the instructions given to the subject to perform fast and errorless.

It should be noticed that the subjects were given no formal preparatory signal such as is often used when testing for reaction time. The present results may be of interest in work studies, when, in serial tasks, a response to a stimulus may always be itself the signal for a preparation to the next stimulus.

Author's address: Dr MARIE-CLAUDE HENNEMANN, Centre d'Etudes Bioclimatiques, 21, rue Becquerel, *67 - Strasbourg* (France)

Medicine and Sport, vol. 6: Biomechanics II, pp. 323–328
(Karger, Basel 1971)

Retention of Simple Motor Response[1]

G. E. STELMACH

University of California, Santa Barbara, Calif.

Introduction

With the mounting evidence that interference plays a major role in verbal short-term memory, emphasis has switched to short-term motor memory to examine whether forgetting in both domains follow the same basic laws [ADAMS and DIJSKSTRA, 1966; POSNER, 1967, and STELMACH and BARBER, 1970]. Little is known about short-term kinesthetic retention; what data are available suggest that short-term motor memory may parallel that of verbal behavior [ASCOLI and SCHMIDT, 1969; STELMACH, 1969]. The present experiments were designed to ask basic questions about short-term kinesthetic retention so that comparisons could be made with verbal memory.

Method

Apparatus. The apparatus used in the present experiments consisted of a free-moving manual lever. A complete description of this lever along with the procedures followed has been given by STELMACH [1969a, b]. Essentially, the blindfolded subject's task was to move a lever to a target position defined by a stop peg; after a variable retention interval, the subject attempted to reproduce the target with the stop-peg removed. The subject's deviations from perfect target reproduction were recorded to the nearest degree. Commands for the subject's responses and estimations were given by the Experimenter.

All retention intervals and sequences of events were controlled by an 8-bank programmed timer. Forty-five undergraduate students at the University of California, Santa Barbara, were used as subjects in each experiment. In the first study target positions, retention intervals and prior activity were within subject variables in a

1 This research and the travel funds for this presentation were supported by grants provided by the Committee on Research, University of California, Santa Barbara, Calif.

3×3 Graeco-Latin square design. The second study partitioned target positions and retention intervals into within-subject variables and response similarity into a between-subject variable in a Latin square arrangement.

Results

The criterion measure for retention was the absolute error from perfect target reproduction in degrees. The means of the absolute errors at recall for the three prior response conditions (0, 2 and 4 response) at three retention intervals (5, 15 and 50 sec) are shown in figure 1. If a subject was in a multiple priors response condition, he made either two or four prior positioning responses about a target (±10°) before he made the criterion response. Each point on the curve represents the errors in reproduction for 15 subjects.

It can be seen in figure 1 that there are definite trends of increasing error with increases in the length of the retention interval. For each prior

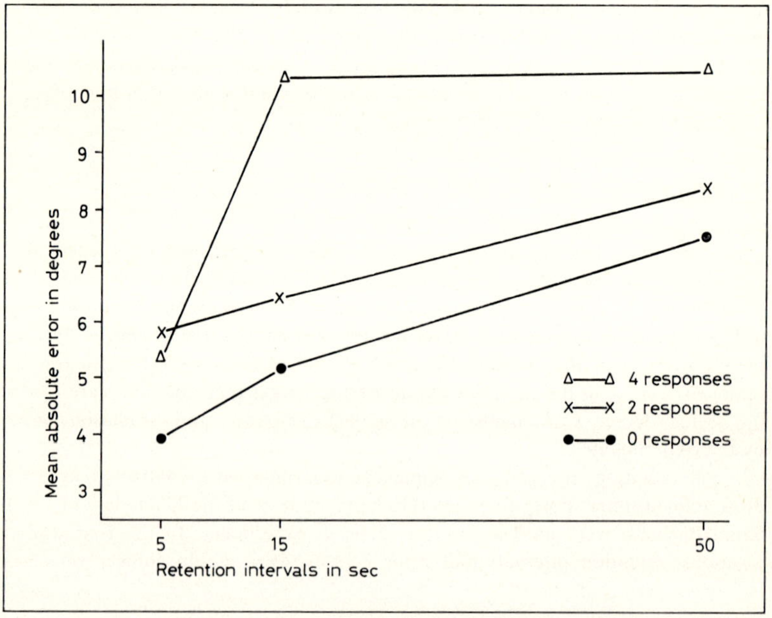

Fig. 1. Mean absolute error of target reproduction plotted as a function of retention intervals for the 0, 2 and 4 prior response conditions.

response condition, there is more forgetting at the 50-sec retention interval than at each of the two preceding intervals. Analysis of variance of the means plotted in figure 1 revealed that the length of the retention interval was a significant source of forgetting: $F\ (2, 84) = 5.95,\ p < 0.05$.

It is also apparent that the number of prior positioning responses affected the accuracy of recall. The magnitude of the forgetting increased with the number of prior responses. An analysis of variance on the response conditions revealed that the variation in errors associated with the number of prior responses was significant: $F\ (2, 84) = 4.42,\ p < 0.05$. In contrast, the amplitude of the movement to the target positions (10, 60 and 90°) failed to yield significant differences.

The fact that increases in the number of prior responses increased the amount of forgetting suggests that PI plays a significant role in short-term motor retention. In the multiple prior response conditions, there are more potential interfering response associations that could interfere proactively with the retention of the target. Theoretically, the loss of information during the retention interval should be a function of the number of prior responses. The results are in agreement with this prediction. Inspection of figure 1 reveals that between the 15- and 50-sec retention intervals the three curves are essentially paralleled. This suggests that proactive interference levels off after 15 sec or perhaps even earlier.

There was also considerable forgetting in the no-prior response condition as a function of the retention interval. This finding makes one cautious of interpreting the results in terms of proactive interference since it is difficult to imagine a source of interference that could produce so much loss of precision.

In figure 2, data are presented from a similar study in which the number of prior responses were kept constant (4) and the similarity of prior responses about a target was varied (± 5, ± 15, and ± 25°).

The basic retention scores were the absolute error in degrees from the target position at recall. The mean absolute error for the 15 subjects within each response-similarity condition is plotted in figure 2 as a function of the retention interval. The striking feature of the results plotted in figure 2 is that the absolute errors at recall are inversely related to the degree of similarity of the prior responses about the target. The errors are progressively larger for the more dissimilar response conditions. An analysis of the mean absolute errors shown in figure 2 revealed the differences between response-similarity conditions were significant: $F\ (2, 42) = 5.45,\ p < 0.05$. An analysis of algebraic errors scores confirmed the above findings that

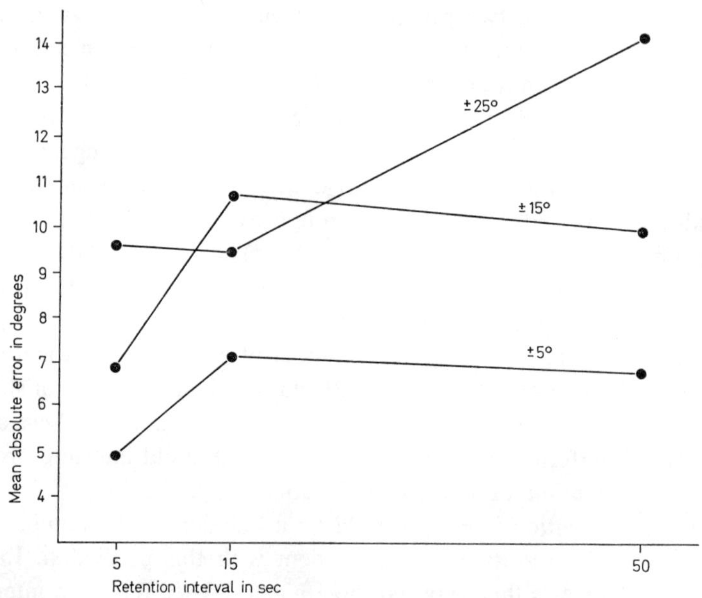

Fig. 2. Mean absolute error of target reproduction plotted as a function of response similarity conditions.

response-similarity is a factor in short-term motor memory: F (2, 42 = 8.02, $p < 0.05$.

Although there is forgetting present in figure 2, an analysis of absolute errors for the retention intervals failed significance: F (2, 84) = 2.89. In contrast, algebraic error revealed that the retention intervals was a significant source of forgetting: F (2, 84) = 4.65, $p < 0.05$. Examination of the main effect target position failed significance.

This study found that short-term retention of a simple motor response was inversely related to the similarity of prior responses about a target. This finding is directly opposite the results usually observed in short-term verbal retention [KEPPEL and UNDERWOOD, 1962; WICKENS, BORN and ALLEN, 1963] and is not consistent with the predictions from the interference theory. The reason for this opposite finding may be in the method of retention measurement. Adequate comparisons of the retention of verbal tasks and motor tasks are beset by many difficulties. Examination of the recall requirement for verbal and motor tasks will reveal that the nature of the retention measurement is quite different. Recall of a verbal item is counted either as correct or incorrect. Therefore, subjects may produce two of the three letters of a nonsense syllable correctly and it is not counted as a correct response, or he may not make a response at all. Much theoretical significance has been attached to this type of analysis; it is on this basis that the two-factor interference theory was first formulated [MELTON and IRWIN, 1940].

In contrast, recall in motor tasks is recorded in degree of accuracy. The retention response is recorded as plus or minus a number of units from perfect reproduction. Since retention scores are not recorded on an all or none basis as they are in verbal behavior, credit is given at recall in terms of degree of accuracy. In conditions where positioning responses are similar, the targets fall within a very restricted range. Consequently, when the subject attempts to recall a target, even if he cannot differentiate between positioning responses or has mis-perceived the target, his recall score in terms of error from perfect responding will be small since all of his responses fell within a small area. Based on the above differences in the measurement of retention between verbal and motor responses, retention scores for motor behavior may not reflect interference in the classical sense. Adequate comparison of the retention of verbal and motor responses must somehow equalize the two types of tasks on this variable. In the future, experimenters examining retention of motor responses should consider using a limited range of tolerance in counting a recall response correct so that retention scores for motor responses will be more comparable to those of verbal behavior.

Conclusion

It may be concluded on the basis of the present data that short-term kinesthetic retention was affected by the number of prior responses. This finding suggests that proactive inhibition effects were present in kinesthetic retention and that verbal and motor short-term memory may be governed by the same set of laws. In addition, it was found that retention was inversely related to the similarity of the prior responses. These results were in direct opposition to the findings usually observed in verbal tasks. Comparisons with verbal responses are difficult to make since the nature of retention measurement is different in verbal tasks.

References

ADAMS, J. A. and DIJKSTRA, S.: Short-term memory for motor responses. J. exp. Psychol. *71:* 314–318 (1966).

ASCOLI, K. M. and SCHMIDT, R. A.: Short-term motor retention. J. mot. Behav. *1:* 29–36 (1969).

KEPPEL, G. and UNDERWOOD, B. J.: Proactive inhibition in short-term retention of individual items. J. ver. Learn. ver. Behav. *1:* 153–161 (1962).

MELTON, A. W. and IRWIN, J. McQ.: The influence of degree of interpolated learning on retroactive inhibition and the overt transfer of specific responses. Amer. J. Psychol. *53:* 173–203 (1940).

POSNER, M. I.: Characteristics of visual and kinesthetic memory codes. J. exp. Psychol. *75:* 103–107 (1967).

STELMACH, G. E.: Prior positioning responses as a factor in short-term retention of a simple motor task. J. exp. Psychol. *81:* 523–526 (1969a). – Short-term retention as a function of response similarity. J. mot. Behav. *1:* 37–44 (1969b).

STELMACH, G. E. and BARBER, J. L.: Interpolated activity in short-term motor memory. Percep. mot. Skills *30:* 231–234 (1970).

WICKENS, D. D.; BORN, D. G., and ALLEN, C. K.: Proactive inhibition and item similarity in short-term memory. J. ver. Learn. ver. Behav. *2:* 444–445 (1963).

Author's address: G. E. STELMACH, Department of Ergonomics and Physical Education, University of California, *Santa Barbara, CA 93106* (USA)

Medicine and Sport, vol. 6: Biomechanics II, pp. 329–330
(Karger, Basel 1971)

Final Remarks

E. JOKL

On behalf of the International Council of Sport and Physical Education of UNESCO, it is my privilege to express the thanks and appreciation of all participants of the Second International Congress on Biomechanics in Eindhoven. Our special gratitude is due to Professor J. F. SCHOUTEN and to Mr. VREDENBREGT, to N.V. Philips' Gloeilampenfabrieken, Eindhoven, to the University of Technology, Eindhoven, to the Institute for Perception Research, Eindhoven, to the Netherlands Sports Federation, The Hague, and to the Netherlands Ministry of Education and Science. Without their generosity this congress would not have been possible.

It behoves me to say something about the categorical relevance of the scientific sessions held during the past days. It has, I think, become evident that biomechanics can and must henceforth be delineated as a subject of its own. It has its own techniques of investigation and its own object of studies: human motion in general and its application to sports, physical education, rehabilitation and orthopedic surgery; by contrast, human engineering and experimental biomechanics are but part of general physiology. Biomechanics cannot be studied in isolation from pure and applied physics, biochemistry, electronics, mathematics, and mechanical engineering. We shall in the future try to focus our activity on biomechanical aspects of sport and physical education upon three complex problems, namely:

(1) the integrative action of the nervous system as far as it designs and controls movements;

(2) afferent and efferent transfer processes which link the central nervous with the musculo-skeletal system, including studies of the muscle-spindle and of electromyography;

(3) the motor act *sensu strictiori,* with special reference to the application and further development of photo-, cinemato- and strobographic as well as mechanical and kinetic methods of investigation.

Exhibitions at the Congress

Scientific Instruments

Almara N. V., Rokin 86, Amsterdam, Netherlands
A. N. R. U. N.V., Wijnhaven 80, Rotterdam, Netherlands
Hewlett-Packard Benelux N. V., de Boelelaan 1043, Amsterdam, Netherlands
Hollinda N. V., Stadhouderslaan 112, The Hague, Netherlands
Kistler Instruments A. G., Eulstrasse 22, Winterthur, Switzerland
Loos & Co. N. V., Transformatorenweg 40, Amsterdam, Netherlands
Gebr. Mijnhart CV, Hoog Kanje 94-106, Zeist, Netherlands
Philips Nederland N. V., Boschdijk 525, Eindhoven, Netherlands
S. E. Laboratories N. V., Vondellaan 128, Papendrecht, Netherlands

Books

Swets & Zeitlinger N. V., Publishers, Keizersgracht 471-487, Amsterdam, Netherlands

Acknowledgements

The Organising Committee gratefully acknowledges its appreciation to the Netherlands Ministry of Education and Science, and the Netherlands Sports Federation in The Hague, for their financial support, without which the 2nd International Seminar on Biomechanics could not have been realized in accordance with the envisaged aim.

The Committee members would also like to express their gratitude to the Institute for Perception Research, the University of Technology, and N. V. Philips' Gloeilampenfabrieken, Eindhoven, Netherlands, for offering their valuable assistance.

Subject Index